Microbe Power
Tomorrow's Revolution

MICROBE POWER
Tomorrow's Revolution

Brian J. Ford

Illustrated by the author

STEIN AND DAY/*Publishers*/New York

First published in the United States of America, 1976
Copyright © 1976 by Brian J. Ford
All rights reserved
Printed in the United States of America
Stein and Day/*Publishers*/Scarborough House,
Briarcliff Manor, N. Y. 10510

Library of Congress Cataloging in Publication Data

Ford, Brian John.
    Microbe power.

    1. Microbiology. I. Title.
QR41.2.F67   576   76-7437
ISBN 0-8128-1936-5

Contents

Illustrations

Microbe Power
Tomorrow's Revolution

Introduction

Most of us kept pets as children: a cat, a dog, or a goldfish. Sometimes these interests established values that are reflected in what we choose to do as adults – and perhaps that explains why I came to write this book, for my favourites were not 'pets' at all, but microbes. To watch a translucent, busy little animal scurrying around its private world under the microscope, seeking out particles that it carefully checks and sifts for food, feeling them with fine finger-like projections and guiding them into its mouth can be as intriguing as watching a kitten at play. Indeed, in many ways it is very much more intriguing, for we know a good deal about the mechanics of a kitten and how it perceives its surroundings: we know very little about the comparable mechanisms that make microbes work.

This essential element of mystery is one of the reasons why microbes have continued to preoccupy my working life. We know much less than we should about the workings of a cell, and particularly about the workings of cell communities. Another reason is the importance of microbes. They are actively involved in almost every sphere of the earth's activities, and with the natural processes of mankind; not just the rotting down of compost or the brewing of beer, but the building of the landscape, the control of pollution and the preservation of health; and we could harness them to feed us, even to mine rare metals that lie trapped within the earth's crust.

Microbes were first written about and studied three centuries ago. For two hundred years after that they were largely ignored by public and specialist opinion, and it is only during the last century – since, say, the 1870s – that widespread interest has been shown in them. Most of that has centred on microbes as the bringers of infection and disability, so we have been left with an instinctive belief that the term 'microbes' is synonymous with *bacteria*, and that 'bacteria' mean *disease*. Does this imply that it was a bad thing to

1

present the germ theory to the public? Not entirely; the notion that there were harmful, transmissible entities in unhygienic situations helped make people more aware of the need for higher standards of public health, and did much to control epidemic outbreaks.

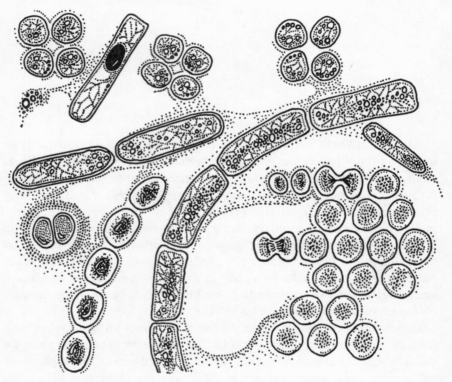

Fig 1: Some bacteria. The two main types of bacteria are rod-shaped and spherical, and often grow in recognisable relationships to each other. The rounded forms can grow in chain formation, in groups of four or eight, sometimes in pairs; or they line up in a regular lattice. The rod-shaped bacilli (one of which is shown forming an internal, dark spore) are often isolated, but can produce very long end-to-end chains.

But in statistical terms, the number of microbes that cause disease is a minute proportion of the species that exist. Many of them confer health on higher organisms, and if we were more aware of the good side of the microbial universe we would discover endless examples of the way microbes help mankind, and many new means of harnessing their vast energies.

Rather than thinking of microbes as essentially *bad*, almost malevolent, let us instead conceive of them as being – in principle – tirelessly energetic workers for *good*. The sceptic might say that thinking about them at all would be a start; everyone is familiar with such specialist sights as the rings around Saturn or the intimacies of a surgical operation, though few are ever given much opportunity to show any interest in the microbes with which our bodies are populated. Most of the public must have looked through a telescope or a pair of binoculars at some time, but how often do people have a microscopist's view of microbes? This hidden universe of single-celled forms of life was here long before we were, and will remain long after our kind has gone; the teeming communities that inhabit your body, your mouth, your fingernails, outnumber countless times over the human population of our planet. Yet to the public they remain unfamiliar.

Strangely enough, this 'blind spot' shown towards microbes extends into the realms of microbiology itself. There were no safety laws controlling the use of microbes for years, although the legislation covering isotopes, drugs, and industrial chemicals was singularly prompt in its appearance. But everyone overlooked the microbe. Proposals to harness useful microbes to produce food and fuel have been tried out in the past, only to fall by the wayside in the fullness of time. There are no insuperable technical or economic reasons for this; usually it was the case of a pioneer trying out a notion that opinions and attitudes of the time could not accept. The examples of a microbe product that did eventually materialize (like penicillin, for example, and the 'miraculous' biological agent in modern washing powders) took far longer to reach practical, commercial reality than one would expect. Even today, a great many microbiologists work routinely with cultures they rarely study under the microscope, so that the research worker tends to be concerned with the products of microbial activity and with the components of cells he can collect and analyse in bulk.

Those who do look at microbes usually restrict themselves to examining a dead, dried, stained smear; or they may observe instead the artefact of shadows cast on the cathode-ray screen of an electron microscope. Few scientists in the speciality settle down to watch the activities and responses of microbes as they go about their business of living.

For many branches of investigation, 'second-generation' evidence is useful. But until we begin to look more closely at the way cells interact, and to study the way they move and respond, we will not unravel the immensely important issues of the way in which cell society is organized. Studying stained permanent preparations tells us next to nothing about cell interaction: it is like trying to study sociology through an examination of corpses.

3

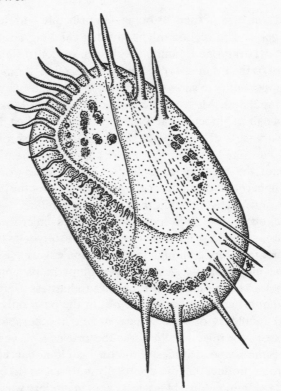

Fig 2: A mouse-like microbe, the pond-dwelling *Euplotes* produces from its single cell a series of projections, with which it scurries actively around, darting forwards and backwards in search of food. There is a controlling centre within the cell which seems to be joined to these appendages, mimicking the way nerves and muscles form in higher animals.

As everyone knows, genetics and the chemistry of life are beginning to become easier to understand through studies of microbes and the way they respond. The so-called 'new biology' arose through discoveries about the way some microbes behave. But what we must do is look further than this: towards the ways in which a new awareness of the microbe universe could alter, extend, and improve our understanding of the future.

Bringing such vitally active organisms to life within the compass of a book is a daunting task; indeed, in any real sense it is an impossibility. There must be enough material in synoptic form in every paragraph to expand into a volume of its own, but even if one cannot be exhaustive at least it is possible to try and aim at being representative of what microbes do, and what they are. I have included some drawings of microbes. These are politely known as 'line-and-stipple studies', but they are really nothing more than sketches of living microbes. A great many teaching books over the years have borrowed diagrams from each other, until the organisms that appear in even the most reputable sources look quite unlike the original. (I have known students who were familiar enough with drawings of a given microbe, to be quite unable to recognize they were looking at it when presented with a live and active specimen under the microscope.) Sometimes it seems that some mythical microbes have been especially invented to populate the pages of school books.

Of course, shading, stippled dots, and lines do not exist in nature as they do on the printed page. The transparent delicacy of a microbe is inimitable,

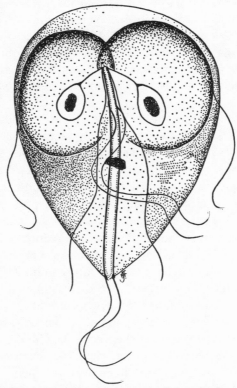

Fig 3:   Giardia with its extraordinary face-like appearance flits around like a falling oak-leaf in a breeze. The microbes in this family are commonly found in the intestinal populations of mice, whilst others live in still water pools. The two 'eyes' are in reality saucer-shaped depressions which the organism uses as a sucker to anchor itself from time to time when it comes to rest.

and drawing them for publication is as much of a distortion as would be a drawing of ripples on a pond, or a diagram of a tear-wet cheek. It takes imagination to put back the glistening, quivering activity which the act of sketching eliminates. But since this has to be a book, and not a film, and because photographs of dead microbes take so much of the vital nature away from the end-result, then 'line-and-stipple studies' are perhaps the best one can offer. The scientific names of microbes have been kept to a minimum, but they inevitably occur none the less. One cannot talk and argue about microbes without knowing what they are called; and in some cases the names are an intriguing comment on the organisms themselves. An example is the bacterium which grows on bread, producing a blood-red pigmentation. It used to occur in churches, and gave rise to many stories of transubstantiation when communion bread had been left out for a few nights. What better name could one imagine for a coloured microbe with such prodigious effects than *Chromobacterium prodigiosum*? After that it is with no little regret that I must add that the bacteria have been given another name, *Serratia marcescens*, which may have some scientific justification, but which is far less worthy. However, the blood-red pigment is still called *prodigiosin*, for all that.

In any event, Latin names should not be seen as forbidding. To gardeners, amateur and professional, they are entirely familiar, and if *Mesembryanthemum* and *Cheiranthus* are good enough for talk over Sunday afternoon tea and sandwiches, then such delicately-named microbes as *Tintinnopsis* and *Vorticella* ought to present no deterrence to anyone.

The spelling of microbes' names brings some problems. The classical diphthong is now disappearing in science, but to many European eyes *Ameba* is an affectation, although *Paramecium* (which was spelled -*oe*- until recently) is now acceptable. I have kept *Amoeba* in this book. American eyebrows may raise a little over the use of the tonne (1000 kilogrammes) instead of the ton (equalling 907 or 1,016 kilogrammes, depending on where you live). I sympathise, although the new term is at least unambiguous. In Europe the ton, as a unit, is on the point of becoming actually unlawful and one cannot doubt that the same will happen in the United States before long.

Whilst in this somewhat apologetic vein, I should say a word to the many colleagues who expected this book to be out a year ago. The reason for the delay was a bout of serious illness – and although we are familiar enough with the propensity some bacteria have for producing illnesses, there is some comfort in the fact that, without the energies of our drug-producing microbe allies, I might not have been here today to tell the tale. None-the-less, much of the scientific background to my book I have explained at lectures over the past few years, and accounts have appeared in international journals published in England, and also in the United States; although this is the first serious attempt I have made to persuade a wider audience of the need to rehabilitate the microbe.

Looking at microbes as allies rather than enemies brings many benefits with it and provides a host of new ways in which we can view our current problems: certainly it brings with it some surprises. It may be, for instance, that the occurrence of microbes on our skin is not something to cause concern, but a supremely important factor in the maintenance of a healthy complexion. If that is so, then regular and thorough washing (that scourge of the small boy) may be an unhealthy thing to do. The energies of microbes in nature are immense, far in excess of anything we can imitate by industrial methods. Microbes can process raw materials at a rate that puts technology in the shade; and their thoroughness and efficiency make the mere machines of men look clumsy by comparison. Here is a clue to a new kind of industry for the future.

Microbes can show us ways to banish pollution, feed the hungry (and feed more healthily the already overweight), and run our future factories without the need to expend fossil energy. Once we tune into the microbe's way of life we can understand ourselves better; we can look again at death and immortality, and find some intriguing possibilities for a reassessment of diseases like cancer, when cell communities run riot.

By writing about microbes without focusing on disease, one discovers endless possibilities for the future. At the philosophical end of the spectrum of ideas lie the ways in which microbes can teach us more about ourselves; while on the practical front they can enable us to turn almost any waste material into fuel and food. To accomplish this we have to hope for a revolution in attitudes that in some ways could be as profound as those that lay behind the germ theory and, in some respects, the industrial revolution itself. Modern society tends to distrust the microbe – and that is the first thing to change.

1 Microbes and our Way of Life

It is no secret that living organisms are living cells: either communities of many cells, in which case the organisms are familiar to us as plants and animals, or else single cells, in which case they are microbes. Are these unicellular forms of life unfamiliar because their ways of working are too specialized for us to understand, or is it their smallness that makes them mysterious? I doubt it: they show us life at its most condensed, and in many ways at its most fundamental and meaningful level. Understand a man and you will learn little about his cells; but understand a cell and you have the biology of man at your fingertips.

This must all seem heady and unrealistic to anyone new to the microbe world – surely, you might think, their influence has been no greater than the effects of countless other natural agencies? Then let us turn to some examples of microbes at work in the world. Consider the white cliffs of Dover, bread and cheese, leather, a pint of bitter ale, or the rounded limestone hills that roll so majestically across the landscape. What more potent symbols could there be of a way of life steeped in tradition, built on a land moulded by the physical processes of geology? Surprisingly enough, these are all examples not of mankind's ingenuity or natural mechanics alone, but of the microbe. Those chalk cliffs, so deep-rooted in the minds of people everywhere as the touchstone of Britain, are the skeletal remains of incalculable numbers of microbes that flourished in the Upper Cretaceous seas, proliferating almost endlessly under a tropical sun one hundred million years ago when the giant dinosaurs held sway.

It may seem almost unbelievable that such diminutive organisms as microbes, each about as small as one could see with the naked eye, could alone and unaided produce the massive stalwart cliffs we see today. But this is an indication of the collective might of microbes working in concert. It is not just an interesting little anecdote to recount, but a pointer towards new forms of technology that – in just this systematic, concerted way – could undertake the most enormous tasks for mankind.

Taken on the basis of a single dwarf cell which divides every hour or two, it is easy to dismiss the whole process as a small-scale phenomenon almost entirely without importance. But think instead of a population of microbes amounting to billions of billions of billions; consider a vast oceanic stretch of them living so densely in the water that it is turbid with their presence. Imagine a million tonnes of microbes dividing, together, into two. After a couple of hours each cell has become two, so a million tonnes of microbes have produced a million tonnes of new life.

Put this way, the massiveness of the proposition becomes clear. In just 240 minutes, 1 million tonnes of microbes, under ideal circumstances, could give rise to 16 million tonnes. Now, that may be juggling with figures, but it is true that we can learn much about natural processes by watching what happens to a single cell, and then scaling up the model. In this way we can sketch out staggering processes of proliferation that mankind can scarcely begin to comprehend, let alone imitate.

How intriguing it is that the largest fossil remains are of the smallest forms of life. Most of our sedimentary rocks are the remains of microbes, fossilized in bulk. Without the microbe, the carboniferous forest ferns would not have partially decayed to give us the carbon-bed we mine as coal. We would not have any limestone for our blast-furnaces either, and even the beds of iron ore so vital to society are largely the remains of iron-oxidizing bacteria. Coal, limestone, and iron ore – the three basic ingredients of an industrial society – have all been bequeathed to us by microbes. Without those ingredients, we would still be struggling for civilization; without the microbe, they would never have existed at all.

Beer, wine, and bread: all three are staple items of Western culture, yet without the intervention of microbes they would not exist either – or not in any recognizable form. The energy of sunlight, captured by the tracery of pike-leaved corn standing in vast tangled acres across prairies and farmland where each plant's leaves are accurately orientated to trap the maximum of light and the minimum of its neighbour's shadow, is stored away in the seed-grains like a battery of natural solar power. And it is on starch-rich grains, replete with energy from the sun – trapped nuclear power, one could say – that man feeds. They are man's fuel.

As man breaks down the molecules produced by the sunlight shining on the wheat, he moves, he breathes, he lives; and the warmth of his living body, like the motive power of his every movement, is the same stored chemical energy that shone as sunlight and was collected in those fields when the corn was gold and green across the acres. The energy debate of the past few years has placed emphasis on solar energy; and houses that can be warmed by collected sunlight, or machines that can turn when the sun shines on them, have attracted much interest. Have we forgotten that mankind himself is a solar-powered machine, or that the human brain is the only solar-powered intelligent computer known to man?

Starch itself is unappetizing, so mankind produces more elaborate food-stuffs by processing the raw grain. The same material can give us, at one end of the scale, a drink – beer; and at the other extreme, a nourishing and digestible food – bread. Why we regard these products as 'natural' is beyond me, for both result from processing of an extremely 'unnatural' nature. Perhaps they seem acceptable to us because they have such an ancient lineage, and have evolved over many thousands of years as man has tried different ways of treating and storing foodstuffs. Traditional they certainly are: yet they represent an advanced form of biological technology. Not only are bread and beer produced from essentially the same raw product, but both are processed by yeast. And it is this microbe which makes us realize that man's contact with the microbe world, and his harnessing of microbes as a source of power, are not ultra-modern at all, but have ancient and well-tried precedents.

Microbes are vital for the manufacture of these food products from the harvested grain; and without microbe activity there would have been no grain in the first place. The plant's supplies of nitrogen and the organic materials it needs for growth derive directly from soil microbes. Even the minute green chloroplasts within the leaf cells, which actually trap the radiating solar energy, may originally have been independent microbes of a sort, enslaved and put to work within the plant for its own benefit.

The age-old practice of rotating crops, so that for one year an area of ground is allowed to grow a crop of clover or some other leguminous plant, was a method of harnessing microbes, too. The nodules which develop on these plants' roots are an infection with *Rhizobium* bacteria, which can take the molecules of nitrogen gas out of the atmosphere and synthesize them into nitrates. To imitate this process of nitrogen fixation on an industrial basis is enormously expensive, and consumes a great deal of energy. Medieval farmers found out how to obtain the same result by using the power of the microbe to fix the nitrogen right there in the soil.

Cheese is another ubiquitous food which has been produced for countless

11

years by microbes. The growth of bacteria that naturally occur in milk – lactobacilli – curdles the milk, so the the solid curds can be separated off and allowed to mature as a firm, fresh cheese. As the bacteria exhaust their supplies of food and water they die, releasing enzymes which split fats and proteins into simpler compounds that are more digestible for man. The energy for all this activity comes from the milk residues themselves, of course, so that as time goes by the calorie value of maturing cheese falls. The result of the microbe's intervention is that the cheese becomes less fattening, more durable, and easier to digest, than the original compounds in milk. The large residue of dead microbes in the cheese is itself a valuable source of body-building protein for the person who eventually eats it.

Microbes have played an important role in giving us our traditional foods and drink, and they have provided us with clothing too. Linen and leather are both products of microbe power. Leather is obtained from animal hide by removing the soft cellular material so that only tough resilient fibres remain. To process leather in this way by purely industrial or mechanical means would be complex and energy-consumptive. But microbes are small enough to get in amongst the structure of the leather, and sufficiently selective to destroy exactly what the tanner wants to eliminate, while leaving the leather fibres untouched. The hides are soaked in pits, and microbes (such as *Bacillus erodiens*) attack them and break down the unwanted tissues. Softer leathers are further processed by drenching, a process in which they are steeped in a mash of fermenting bran or – in some processes – cow-dung.

This may seem a repugnant way to treat the finest and most supple of leathers, but it is nothing more than the best means of obtaining active communities of microbes that can digest the unwanted components of the hide and provide us with the pure, refined product we want. All the various mashes and bubbling tubs of fermenting brew, the timing, the temperature, and the traditional ingredients used by the leather tanner are the end-result of centuries of experience. They represent the perfection of microbe technology, and man knew about it long before he was taught that germs cause disease. Incidentally, the addition of tanning compounds (such as oak-gall) at the end of the process is simply a way of rendering the leather non-degradable during its period of use – mankind's method of finding out how to switch off the power when required. So leather production is yet another example of microbes working efficiently for man. Why then is it that the process is hardly ever referred to in textbooks on microbiology?

Hemp and flax are age-old examples of plant fibres put to use by mankind.

In the growing plant, the fibres we find so useful are bound tightly to the rest of the plant structure by pectin and related glue-like substances. Mechanical separation is possible, but inefficient. However, if the plant stems are cut and steeped under water, microbes can get to work with predictable ease. First to proliferate are the microbes that – like man – breathe oxygen, the aerobic microbes. They rapidly decompose the softer tissues, but use up the supplies of dissolved oxygen at the same time. From then on, butyric acid bacteria begin a methodical degradation of the remaining cellular residues until eventually only the tough fibres remain. The resulting threads can be spun, and then made into rope or twine, or woven into fabric. Without the involvement of the microbe (which played a vital role in the soil when the crop first grew, of course) throughout the process, none of this would have been feasible. And though one could design a machine to take over parts of the procedure, the brute force of machinery is no match for the precise and systematic activities of the microbe.

What other examples of microbe processing can we cite? Vinegar is an obvious candidate; originally it was an accidental end-product of wine and beer fermentation, when too much oxygen entered the brew and aerobic microbes appeared. They oxidized the alcohol to acetic acid. When vinegar was to be produced intentionally, it was made by running an alcoholic liquid through barrels of wood chips or twigs – ideal for the microbes, as it allowed ample oxygen to percolate through the mix.

There are even microbe cosmetics. The much-prized ingredient of the most expensive perfume – ambergris – is the decomposed remains of putrefying cuttle-fish, broken down by the microbes of decay. The algae and fungi that together make up colonies of lichens have given us *mousse de chêne*, which is prized for its scent, and dried lichens were used as wig-powder in the past. Certain microbes from soil have been shown to synthesize useful molecular complexes which, though not perfumed themselves, are a valuable raw material for processing into cosmetics.

Microbes may seem an unlikely source for exquisite perfumes. But in some parts of the world they have long been used as an item of diet, and here I am not thinking of foods that are processed by the energies of microbe activity, but microbes themselves being eaten. The ground around Mount Asama in Japan reportedly supports the growth of pure colonies of bacteria that are dug out and used to make a dish known to gourmets as Broth of Tengu, and the Chinese (like the people who live around Lake Tchad – see p. 54) eat preparations of algae. The Chinese product is To fa Tsai, 'hair crop', and it consists of a soup made from the blue-green alga *Nostoc*.

Though that is a delicacy, the crop of algae harvested at Lake Tchad is a staple item of diet.

So man has already established traditional uses for microbes. The processes are unique in that they do not require industrial energy, or the consumption of large amounts of fuel; and they are potentially non-polluting. They could, were we to overcome our inbuilt dislike of microbes, and our conviction – learnt from childhood – that they are likely to harm us, produce a host of valuable new products; and they could enable mankind to find a new form of biological energy at a time when we desperately need new thinking in this field.

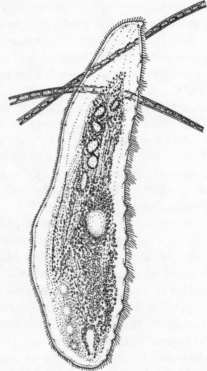

Fig 4: A voracious feeder, *Loxophyllum* possesses a long slit-like aperture through which it can ingest other organisms larger than itself. Much of the cell is transparent and very thin. Note how clearly the two filaments of algae over which this example is passing, can be seen through its body. The microbe is propelled by the steady flickering movement of fine hair-like cilia which literally row it along.

The argument I want to put forward is this: if microbes can be eaten, and if they have in the earth's past proliferated in such vast amounts, should we not seek to use them more widely than ever before as a source of food in a hungry world? If yeast has given us such diverse products as bread and beer, what other microbe processes could give us similar processed foods that would be more realistically aimed at providing nourishing foodstuffs for all nations, rather than the energy-consumptive luxury diets that are killing off millions of overfed people in the West? As we obtain alcohol from *Saccharomyces* and acetic acid from *Acetobacter*, what other useful products could other species of microbe provide for us? If some microbes can convert corn-steep liquor – a waste product – into penicillin, or dead cuttle-fish into perfume, what more vital industrial processes might still unheard-of microbes carry out for us, turning surplus materials into products we may need to survive? Finally, what new insights into ourselves, and what new interpretations of health, can the microbe provide? A systematic investigation of the way microbes work in nature, rather than in the test-tube, could enable us to find new ways of protecting man, his animals, and his crops from disease or disability.

In these fields, and in many more, the energies of microbe power must now be harnessed. The principle is hardly new since for millennia we have produced so many staple items of our diet and our culture with the aid of microbes and many large-scale production ideas have come near to fruition in the past, only to fade from view before their true benefits could materialize. Since the beginning of this century various experiments have shown how protein could be provided for us by the mass culture of microbes, for example, and during the Second World War some limited practical use was found for the idea. Indeed it is not many years ago that a brand of petrol (sold in Britain under the trade name Discol) was on sale, containing a percentage of microbe-produced alcohol. How ironic it is that the era of the oil shortage should coincide with an absence of a fuel like this from the pumps.

But these pilot experiments, from the fuelling of automobiles with microbe alcohol to the production of cattle-cake by the fermentation of oil waste, are not the answer. The much-publicized chicken-shit engine which was claimed to be running on excrement was actually fuelled by methane produced by microbes in the fermenting brew – and I think it would have created a far more vivid impression, and a more accurate one, if we had been told that you could run a motor-car on microbe energy like this.

However, in a climate of opinion which is not generally interested in microbes, in an age in which the public (and the majority of scientists) continue to hold false impressions of the microbe, and in which the importance of microbial activity in the workings of our planet has yet to be appreciated, these experiments are not the answer. What we need first is a revolution in thought – a reversal in our attitudes to the world of micro-organisms. For a century we have feared them as the agents of disease, and the bringers of death. Now we have to look further, to see a wider view; and now is the time to see that the microbe can provide a new way to live.

2 Microbe Power

Every living cell is a controlled powerhouse of energy. We may consider a single microbe, or a cell from within the complex community of separate entities that together comprise a human being, but each one produces energy from the breakdown of food materials. The result is a rate of production no man-made factory could emulate, and it could provide us with an unlimited prospect for biological technology. Our model in history has been the yeast of fermentation, *Saccharomyces cerevisiae*. In still wine, the yeast's by-product of alcohol is accumulated as the carbon dioxide bubbles away; in bread it is the carbon dioxide which is trapped and which causes the dough to rise, while the alcohol evaporates. In brewing we want the alcohol: in baking the gas. Yet both these products are wastes to the yeast. It is not an overstatement to say that alcohol and the carbon dioxide are the excrement and the exhaled breath of the living microbe. Indeed, both products are poisons to the living cell which produced them, and the accumulation of either in excess kills the microbe from which they came.

To the yeast, in its confined world of a fermentation vat, these products are pollution. In nature they would either be diffused, or they would be utilized by other organisms as raw materials for growth. But trapped in the wine-maker's container they accumulate until the alcohol poisons its producer. So the yeast microbe dies, suffocated in its own wastes. But to man, the by-products are potentially valuable. The alcohol is not only used as an agent of commerce and an intoxicant, but it is an important industrial raw material and a useful solvent.

Of course, we can produce alcohol by industrial, chemical methods. But the application of external energy to bring about the reactions is wasteful in the extreme, when all the time there is a microbe available that can do the same conversion by using the molecular energy within the substrate itself.

Microbe Power

The yeast converts carbohydrate molecules to alcohol and carbon dioxide, and the energy released by that reaction is all the cell requires to fuel itself. The lesson we should learn from this is twofold:

(1) It is wasteful to apply external energy to any reaction that is capable of fuelling itself.

(2) It is wasteful to utilize industrial processes for any reaction that microbes can carry out.

The exothermic, energy-giving reactions of life are capable of being harnessed for all manner of industrial tasks and it is now, with the demands for energy approaching extreme limits, that we should begin to investigate them. We have become familiar with the traditional method of obtaining acetic acid (the oxidation of alcohol by *Acetobacter* bacteria to produce

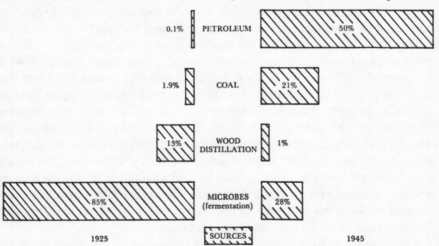

Diag 1: Chemicals for industry: two decades of change. The easy availability of petrochemicals led mankind away from the more long-term prospect of power from microbes. It is a trend that could now be reversed.

vinegar). What is less widely known is that most acetic acid nowadays is made by industrial processing instead. Raw materials such as acetaldehyde or butane gas are oxidized in the presence of a catalyst, which greatly increases the rate of throughput. One wonders why such efforts should have been put into finding an energy-consumptive artificial alternative for a chemical conversion already capable of being carried out by microbes evolved for the task.

Microbes have already won the battle over the production of citric acid, which is widely used in the manufacture of fruit drinks. Citrus fruits produce this material, of course, and it provides them with their characteristically sharp-yet-sweet taste. Some 60,000 tonnes of citric acid are produced each year in Europe alone, much of it for the fruit-drink industry. Considerable efforts have gone into finding an industrial process which could achieve these rates, but they have failed. Instead, citric acid is produced by nothing more glamorous than a mildew, the fungus *Aspergillus niger*, fermenting a growth medium made with molasses (which would otherwise be a waste product). The fungus produces a felty mass of hyphae, converting the waste into a solution containing citric acid. Eventually, if left to itself, the degradation would continue until carbon dioxide and water were all that remained, but the use of techniques to block the reaction allows us to obtain citric acid in bulk.

Here is a perfect example of a self-fuelling reaction, carried out by the living energies of a microbe. Yet man continues to try out alternative methods that might enable him to imitate the same process artificially. The fact that he fails shows how efficient the processes of living organisms are; but the fact that we continue to look at all tells us even more about our blindness to reality.

Gluconic acid is another example of an important chemical raw material produced for us by the microbe. Salts of this acid – gluconates – are widely used in medicine, and have a host of specialized chemical applications; and the most effective way of obtaining the compounds is fermentation. In the paint and plastics industries you will find many applications for itaconic acid (a common ingredient of adhesives). Many alternative pathways for its synthesis in the laboratory have been tried out, with little success; and in this case too the most widely used source of production is the microbe.

Lactic acid, on the other hand, is produced by industrial *and* microbial methods, depending on where you look. In England and Holland the microbe is used to produce lactic acid (we have referred to the *Lactobacillus* in cheese manufacture, on p. 12), whereas in France and the United States the industrial route is preferred. Lactic acid, however, exists in two forms that are chemically identical, but which can be molecular mirror images of each other. Only the 'left-handed' L-form can be utilized by living organisms, whereas the industrial process produces equal amounts of the L-form and the D-form. If biological compatibility is an important aim then it would be preferable to utilize the natural production of lactic acid by the microbe, since the biological approach of cell technology provides the pure L-form, and that alone.

Microbe Power

Once again our quest for industrial substitutes for natural processes has given us an alternative that may not be as good as the original, even in terms of end-product quality; and when we consider the energy consumed by the industrial version of the process it looks even worse. So far the production of lactic acid by microbes is run on a batch-process basis. But there is no reason why we should not derive a continuous-process substitute, and if this came about we would have a clean, efficient, silent, and reliable means of producing L-lactic acid for industry – and this continuous process is not a new idea, either. The trickling of alcoholic liquor over twigs and chippings in the production of vinegar dating from the middle ages is continuous processing of exactly this kind.

Our growing awareness of the need for a well-balanced diet has greatly increased the demand for supplements – vitamins and the like. In our industrial processes, we have developed methods of producing vitamin B_2 (riboflavin) on a mass-production scale. But it is perfectly possible to use microbes to produce it for us instead. Many microbes produce an excess of the vitamin, amounting to quantities far greater than they could ever utilize themselves. At present, the only significant use for microbe-produced vitamin B_2 is in animal feeds. But just let us look at the figures:

TABLE 1: VITAMIN B_2 PRODUCTION BY MICROBE CULTURES

Amount of riboflavin needed in a culture medium	= ½ mg per litre
Amount of riboflavin produced by a typical microbe	= 10 mg per litre
Amount of riboflavin produced by a yeast	= 100 mg per litre
Amount of riboflavin produced by B_2 microbes*	= 5,000 mg per litre

They reveal an astonishing level of over-production, for the B_2 microbes produce ten thousand times as much vitamin as they require themselves. The usual method by which we harness the over-production of the B vitamins by microbes is by chewing yeast tablets. What we forget is that yeasts are capable of producing two hundred times as much riboflavin as they need, so that – instead of consuming the dried remains of their cells – we should be availing ourselves of the vast amounts of the vitamin they produced when they were growing. We could obtain considerable yields of vitamins without expenditure of energy, if only microbes were used as the productive machinery rather than the unsubtle and crude resources of a factory.

One of the most widely known of the microbial food additives is MSG,

---

Ashyba gossypii and *Eremothecium ashybii.*

20

monosodium glutamate, or seasoning salt. The glutamic acid fermentation from which it is derived provides 90 per cent of the world total – only 10 per cent is produced by chemical means, amounting to some 40 million kilogrammes per annum. Though MSG has no strong taste itself, it enhances flavour in savoury foods and is a common ingredient of convenience foodstuffs. The fermentation was first harnessed in Japan, and it has been found that the microbe *Brevibacterium* which is used in the process greatly increases its rate of output if penicillin is added to the growth medium. Furthermore, the organism can use waste carbohydrates as a substrate for growth. So this much used food additive could be produced from discarded by-products.

Like so many potentially useful agents, MSG has been grossly over-used, and cases have occured in which people have reacted to an overdose. The condition is known as Kwok's disease (after its discoverer) or more graphically as Chinese Restaurant Syndrome, and as a result MSG has been restricted in parts of North America. This is perhaps a pity: it can be a useful material, in sensible applications. But the glutamic acid product is not only used to make MSG: it is useful as an important amino acid in its own right. Many foodstuffs are deficient in this particular substance, and it can be added to them to help provide full nutritive value.

The first bacterium to be studied as a commercial producer of glutamic acid was *Corynebacterium glutamicum*, a relative of the diphtheria organism. It has since thrown up a mutant form which produces a large surplus of lysine, an amino acid vital to humans but missing from cereal protein. The availability of lysine would make it possible to enrich many kinds of food presently considered unsuitable for general use, and the microbe can convert the acetic acid in vinegar into lysine with high efficiency. The synthetic production of lysine on an industrial basis is costly in the extreme, but when the microbe undertakes the process the cost of the final product can be considerably less than £1 or $2 per kilogramme. Threonine is another important amino acid which we have in the past obtained by factory methods. A strain of *Escherichia coli* isolated from human faeces has now been found to produce large amounts of threonine, and though the process is relatively new it seems likely that the cost of production could be more than halved if these microbes were used.

Everyone eats the products of microbial metabolism, and the majority of us consume cereal crops which have been – or could be – fortified by vitamins that might best be produced by the microbe. But we do not notice they are there, and the developed countries rarely hear about deficiency diseases even when – as with scurvy, found in vitamin-C-deficient old people – they still occur. By contrast, everyone has heard of the 'washday miracle': the biological washing powders that degrade protein ('dissolve understains' as they coyly say) as nothing else can. Every housewife is aware that it is enzymes in the powder which carry out this process for her. She is less likely to know that they have been produced by microbes.

These enzymes are digestive compounds that dissolve proteins, and are extracted from cultures of the bacterium *Bacillus subtilis*, a common rod-shaped organism found in soil and fresh water. The bacteria are cultured in bulk tanks, and the enzymes they produce are an essential part of the microbe's personal equipment, for they play a vital role in attacking and breaking down the proteins on which it lives. But when extracted, purified, and incorporated into a washing powder they can attack blood, grass-stains, mucus, and any other protein-bound stain that will not yield to normal washing.

The over-use of these enzymes has been criticized (and wisely criticized) too. As so often happens, an industry keen to exploit any new idea to the utmost has saturated the market with miracle washing preparations. It is undeniable that enzymes are enormously helpful, when used correctly; but in some areas, where a backlash has followed careless and extravagant usage, there have been local bans on enzymes altogether. One hopes that, in a more responsible climate of opinion, such extreme controls will cease to be necessary. Life without biological washing powders for most British women would seem harder, that is for sure.

Dissolving proteins is something which microbes are well fitted to do, and which chemistry has been unable to imitate successfully. But other molecules can be split, too, and there are several examples of the way that the microbe can undertake complex chemistry of a kind that it is impossible to imitate by conventional laboratory methods. The lessons we can learn from these examples point the way to a future for cell technology which is almost limitless.

Perhaps the best example is also the first. It was in the late 1940s that an American biochemist, Dr L. Sarett, succeeded in synthesizing the molecule of cortisone. At that time, no one realized that cortisone had important uses as a drug, and the synthesis Sarett undertook was an elaborately successful piece of biochemical juggling. He began with the distantly related deoxy-cholic acid and then painstakingly proceeded through thirty-two successive stages before arriving at the target molecule. The final yield was less than 0.2 per cent of the amounts he started with – 998 parts out of 1,000 were wasted, which is not uncommon in biochemical synthesis. In 1949 the power of cortisone as a drug was discovered. Since then a range of related drugs has been developed and, for all the progress made by Sarett in the laboratory, most of the important processes that have made these developments possible have been undertaken by microbes, not human chemists. The Sarett synthesis would have made supplies of cortisone from non-human sources into an expensive commodity: but in 1950 a team of research workers at the Upjohn Laboratories announced a biological answer to the problem.

They used a microbe – the fungus *Rhizopus arrhizus* – to carry out the conversion, rather than relying on biochemical processing. This suddenly made it seem worth while looking for other microbial species that could

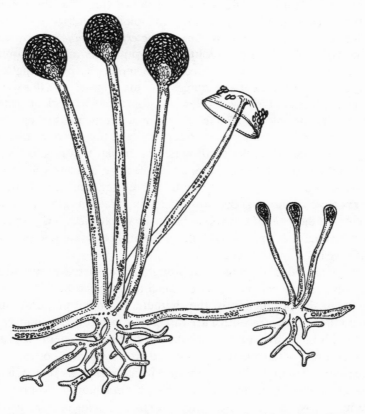

Fig 5: The pin-mould *Rhizopus* can sometimes be found on discarded scraps of bread. The spores, which form in rounded sporangia, are discharged explosively when the spore-case suddenly collapses (an example can be seen in the figure). Other species of *Rhizopus* undertake chemical reactions that are of industrial importance.

work on molecules and produce end-products that chemists could not hope to make themselves. Cortisone – for all its uses in treating rheumatism and related conditions – had serious side-effects. In those early days it was isolated from many sources (principally the bile of cattle) and in a heady, optimistic atmosphere it was given to many patients who were otherwise without hope. Often their bones lost calcium, causing osteoporosis and the risk of fracture; the patients tended to become moon-faced, fat and hairy; and spots and blemishes appeared on their skin. They tended to retain salt. So – although cortisone was a useful drug – it was, as it stood, not the kind of compound one could administer as often as it was indicated.

Modifying the drug chemically, or synthesizing similar compounds, was likely to be a daunting task. Sarett had shown that. But the prospect for microbe processing encouraged others to look for a biological answer – and they found it in the bacterium *Corynebacterium simplex*, another of the innocuous organisms in the diphtheria group. This microbe could bring about a strategic change in the cortisone molecule, producing a similar compound which had clear advantages over the original: and this was named prednisone. If the microbe was given hydrocortisone to work on instead, the result was prednisolone. Unlike the original drugs, neither of these new compounds produced severe salt-retention; and they were as much as five times more potent than the cortisones. This meant that dosages were kept lower, and the possibility of side effects became smaller still. These two steroids are now widely used. They have brought considerable benefits to sufferers from diseases as disparate as leukaemia and arthritis; indeed the full extent of their benefits is still to be finally unravelled. It is just as well we looked for microbes to undertake the production task, for without them this would have been unthinkable.

The contraceptive pill, which has brought about such widespread changes in attitudes and moral values, owes much to the microbe too. Several key reactions in the production of the synthetic hormone preparations which make up the oral contraceptive are difficult or impossible to imitate chemically. So this is one avenue in which the production of a chemical product by the intervention of the microbe has altered even our sexual codes!

The future of microbe processing, of cell technology, as I call it, is likely to be of considerable importance to the pharmacologist. Many microbes (including most notably the fungi *Botrytis, Thielavia, Curvularia*, and *Sporomia*) have a strong tendency to oxidize molecules in the culture medium. They have already been used to tailor molecules at will, such as the oxidation of the antibiotic tetracycline to oxytetracycline. We can only guess what new effects we may obtain by selectively altering the structure of drugs already in existence through the activities of microbes – and no one can predict the new chemical products that microbes have yet to reveal to us.

And what of pesticides? Mankind will always need some means of controlling pests, and the desire of some people to look for an era when crops were grown so 'naturally' that pesticides were entirely unnecessary is unrealistic, since the growing of crops is itself unnatural. Under primitive conditions, when mankind has not intervened, crop plants such as cereals and cabbages grow interspersed with other species as part of a complex community of plants and animals. Because of the dispersed distribution of wild plants, pest species show adaptations that enable them to survive long enough, or in sufficient numbers, to get from host to host. They often produce large numbers of progeny, or exhibit parthenogenesis (the phenomenon of virgin birth, in which a female reproduces without fertilization), so that they stand a chance of survival in spite of the difficulties attached to finding a host on which to feed.

Once agriculture arrived, this propensity showed that it could result in devastation. A pest that infested one plant in a field found it had thousands of identical plants all around it – and so the ability to survive under difficult circumstances became, once man had discovered how to farm, the origin of widespread outbreaks of disease. This is why we have needed to dress seeds, to spray crops, to dust growing plants. But our instinctive tendency to look for a chemical or industrial answer has led us to overlook the many microbes that could be used to control pests. Some of the answers we have found show industrial biochemistry in its best light. DDT, though widely condemned in recent years, has been instrumental in eradicting insect pests and disease-carrying species: an important example of a public health measure with international ramifications. It has done this without significant risks to mankind. That is, in itself, an achievement that it would be wrong to under-estimate.

But DDT did have its hidden side; its bad face. As we now know it accumulates through the successive predators that feed on treated grain – from small birds up to the falcons – and has endangered the very survival of some species. So even this most safe of insecticides has been widely banned and withdrawn from circulation,* and questions are still being raised about the safety of the other chemical control agents now in use.

* One hopes we have not over-reacted. I see that some insecticidal preparations are now being made with nerve poisons more potent and more hazardous than DDT used to be.

Microbe Power

Microbes, however, could provide us with a range of new insecticidal preparations that are harmless to man, ecologically compatible, highly efficient, and selective. The concept of biological control is itself far from new. If you have a dog to keep cats away, or a cat to control mice, then you already have practical experience of biological control. In some instances, microbes have been used in the past to control pests. But when we assemble some of the examples it seems strange that the avenues for microbe insecticides have not been more systematically explored.

Many virus diseases of insects are known, perhaps four or five hundred altogether, and (though viruses are not living cells, and therefore they are not true microbes – see p. 157) many of these could be viewed as candidates for consideration as biological control agents. A large proportion of insect viruses are laid down in the host cell in a protein matrix, forming crystalline bodies that can be separated out. The protein holding the virus particles acts as a protection for them, so that they are only released when one of these particles has been taken in by a new host and the protein 'capsule' has been digested away to free the virus once more. These 'crystals' can be divided into two main groups of insect virus diseases: the *granuloses* in which crystals form, each around a single virus particle, and the *polyhedroses* where many-sided crystals develop, each containing hundreds or thousands of virus particles.

It is a polyhedrosis virus which provides the classic example of control. The pest was the sawfly, which used to devastate large areas of spruce forest in North America. The fly was accidentally introduced from Europe, where it had existed more or less in equilibrium with its host tree and did not cause significant outbreaks. But when it reached America, it wreaked havoc among spruce trees unprepared for the onslaught. It was then that a natural disease of sawflies was brought in to control the outbreak. Large cultures of the sawfly larvae were infected with a polyhedrosis virus. Ten larvae reduced to fragments and mixed with one gallon of water supplied enough virus to treat an acre of young trees. The idea was introduced to infested areas of Canada during the 1930s, and had the effect of drastically reducing the sawfly population, and since then experiments have suggested that other species of sawfly can be controlled in the same way.

There is much in common between the way that sawflies spread rapidly through the dense spruce forests and the manner in which the polyhedrosis spreads through the sawfly community. Given a new 'virgin area' to colonize, or an unusually compact suspectible community through which to spread, any new organism is likely to proliferate until it gets out of hand. A similar overwhelming situation arose after the prickly pear cactus, *Opuntia*, was introduced into Queensland and spread like wildfire. But it is as well to remember that what is essentially 'good' or 'bad' is a distinction drawn by man, and not by nature. There is nothing inherently beneficial about the spreading of a contagious disease – if we choose to select it as an ally because it reduces the numbers of a pest that adversely affects us, then that is only because of our own choice of criteria. I dare say that the agent introduced to control the Queensland cactus explosion – *Cactoblastis* – is regarded as a tremendous pest in the prickly pear fraternity, for all the enthusiasm with which man regards it as a worker for the common good. The lesson is that the implications of control have to be borne in mind, and control itself should not be used indiscriminately.

Myxomatosis and its use to control rabbits provides an example of the ill-considered application of a virus to control a pest. It began with Giuseppi Sanarelli, a hygienist who worked in Uruguay around the turn of the century. In 1896 he introduced European rabbits into the experimental animal house of his new Institute for Hygiene in Montevideo. He intended to use them in the preparation of therapeutic serum, but noticed that some of them began to show signs of disease. Irregular pustules developed, and large nodular tumours began to spread across the animals' heads. The disease proved to be highly contagious, and the rabbits were rapidly incapacitated by its hideously disfiguring effects. Nothing like it had been recorded before, and he coined the term myxomatosis for it. We have since realized that the disease is caused by a virus which was widespread among the rabbit population native to Uruguay but which was new to the European rabbits. It was the classic situation of a virulent pathogen in a new, virgin population that was unaccustomed to its effects.

In Australia the European rabbit had already become a serious pest. From the first few rabbits taken over as pets, to decorate the landscape, they had – by exactly the same invasive process – taken over the bulk of the grasslands and were causing enormous losses to crops. Australians on the look-out for a means of eradicating rabbits felt that the myxomatosis infection might be just what was needed and so, in 1926, samples of infectious serum were taken to Australia to be inoculated into rabbits, which were then released into the wild.

But nothing happened. Several other attempts were made over the following years, but none of them succeeded in establishing an epidemic of the kind the Australian authorities were looking for. Not until 1950 did the attempts succeed. The weather at that time was particularly favourable for the development of mosquitoes, and it was found that they were the principle means of transmission. Myxomatosis is contagious, true; but outbreaks are slow and localized. In Australia the widespread epidemic came with high populations of blood-sucking insect vectors. More than 80 per cent of the rabbits in South East Australia were wiped out in the first wave of the epidemic, and in some areas millions of them were found piled high against rabbit-proof fencing.

Two years later the virus was introduced into France. It was high summer once again, and mosquitoes spread the infection rapidly. It moved in a broad front across the whole of western Europe and in the following year appreared in south-eastern England. Mosquitoes are not as common in England as they are in Europe, and it was the rabbit-flea which took over the task of transmission. By the end of 1955 it was estimated that 90 per cent of the rabbits in Britain had been killed. Their emaciated, grotesquely disfigured corpses were a familiar sight on the roadside. Since that time the disease has remained present in those rabbit populations in epidemic form. The effect of natural selection has been to produce rabbits that are more resistant to the disease, and animals blinded by it may still survive and feed normally. They do not burrow in the way they used to, however. Now they often live on the surface of the ground, like hares.

Rabbits in areas where myxomatosis is endemic can apparently breed even when disfigured by the disease, and blindness does not prevent them from surviving for years. So myxomatosis does not eliminate rabbits entirely: and, for as long as they survive, they can continue to feed rapaciously and to proliferate. Populations of rabbits now increasingly immune to the disease may be on the way to developing tolerance, like the Uruguayan rabbit populations, which now develop nothing worse than a small swelling on the site of a bite from a myxomatosis-carrying insect.

So this virus disease has produced an ineffective form of control, partly because the transmission of the virus – by insect vectors – was poorly understood, and insufficient knowledge was amassed on the natural history of the virus itself. It is also cruel, for the virus acts more by disfigurement than by rapidly incapacitating and killing the rabbit through interference with some vital physiological mechanism. The animals therefore remain able to feed and to breed while disabled, so, far from being eliminated, they multiply. And finally, of course, there is the known ability of rabbits to evolve resistance to myxomatosis.

Viruses as agents of pest control have to be more carefully selected in future. Though most viruses are characteristically host-specific – which is to say they only infect one species – others are not so choosy. Thus the polyhedrosis virus of silkworms (a serious disease which has important commercial consequences) has been found to be equally capable of infecting the waxmoth, which is itself a pest. In this way a pathogen that has caused economic difficulties in the past may be diverted to control an 'enemy' of mankind instead. But the ability of a virus to infect more than one host species introduces a note of caution into the argument. We do not know how widespread this proclivity is, and electron microscope studies of some of the viruses from insects show that they have a superficial appearance similar to that of some viruses which infect mammals. There is no certainty that morphological similarity implies an ability to cross-infect, but in a relatively early state of knowledge we would have to move carefully before adopting virus control agents as pesticides.

However, the use of restriction enzymes, which can enable us to transplant selected sections of genetic material into virus particles, could open up an interesting way of producing viruses with precisely predictable characteristics. So it may be that the long-term prospect is to be the construction of synthetic viruses that could produce a clearly defined effect in the host, and the development of pesticides of this kind would represent an advance of limitless significance.

Bacteria, as true microbes (see Chapter 7), pose different issues. There are fewer bacteria than viruses known to cause diseases in insects, and which might have the ability to act as microbe insecticidal agents. In contrast to the four or five hundred insect viruses known, there are approximately one hundred species of bacteria in the same category. But there are many ways in which bacteria could be used as insect control agents. Viruses commandeer the cell machinery of the host, and genetically instruct it to synthesize new virus particles, which makes it necessary for the machinery to exist in a suitable form. So viruses (even if they may not turn out to be as host-specific as we used to think) have an inbuilt tendency to be restricted in their choice of host. Bacteria, on the other hand, metabolize proteins and other foodstuffs and are clearly likely to be able to utilize foodstuffs of differing kinds as a result. This does not mean that a bacterium which causes boils will do as well infecting the cells of a cabbage, but it does imply that a microbe which metabolizes protein may exist as successfully in a caterpillar as in an ape, or in a wood-boring beetle as well as in the soil. For this reason we are not confined to using only microbes that naturally cause diseases, and some bacteria that do not ordinarily infect insects can be put to that task most successfully.

Microbe Power

Popillia japonica is a Japanese beetle which damages crops and lawn grasses, and which produced widespread outbreaks when for the first time it reached the 'virgin areas' of Canada and the north-western states of the USA. The natural pathogen of this beetle is a microbe, *Bacillus popilliae*; and it has been used in a successful demonstration of the power of microbe pest control. The organism is a rod-shaped bacterium which produces the milky disease of the beetle larvae. The bacterium produces spores which can resist drying, and which can survive for long periods in vegetation.

How does the microbe produce its effects? Spores of *Bacillus popilliae* are eaten by the larva along with its normal diet of vegetation. They can resist the normal digestive processes of the larva, and in due course germinate and release the rod-shaped bacteria which produce the disease. They penetrate the gut wall into the normally transluscent haemolymph – the blood-like fluid in the bodies of these insects – and proliferate on a massive scale. So great do the numbers of the bacteria become, that the larva becomes white with them and develops the milky appearance which gave the disease its name. The microbes act as a severe metabolic load on the larva; they obstruct the free circulation of its vital haemolymph; and the host dies in consequence.

The spore is the vital link in the chain, but for some reason the bacteria do not readily produce spores when they are grown in laboratory culture. For this reason, the large-scale production of *Bacillus popilliae* is undertaken by harvesting infected larvae. Large numbers of them are produced in cages, and are artificially infected with spores of the microbe. After the disease has developed, the larvae are dried and crushed to a powder which releases the spores. An ounce or 28 grams of the commercial preparation contains some 3,000 million spores – approximately as many individual organisms as the entire human population of the earth.

Another spore-forming microbe, *Bacillus thuringiensis*, has provided an effective answer to some insect pests although it is not a natural pathogen to them. Once it has been artificially introduced into an insect community it produces widespread mortality, but the effects are not maintained and no long-term endemic situation results. The microbes have to be readministered to a given area annually for protection to be maintained. This species was first isolated in Japan in 1902, where it was found to be causing a disease of silkworms. Like *Bacillus popilliae* it can be reared on artificial media, but in this case spores form readily without the intervention of the natural host. It is relatively easy to mass-produce the microbe, induce it to spore on the grand scale, and then harvest the infective end-product. Hundreds of tonnes of this organism are produced for sale in the United States each year, where it is registered for use on scores of plants for protection against two dozen different pest species. It has been successfully employed against tobacco and tomato worm, many species of moth, and the cotton boll-worm among others.

Microbe pesticides have, as a whole, many advantages over chemical control agents. There is no risk of cumulative poisoning, or of the chronic toxicity that has been associated with some insecticides and also with dressings containing heavy metals such as lead. There is no addition of chemical pollutants to natural ecological systems. Microbes will not persist in soil to the detriment of later crops, as happens with many chemical pesticides; unlike chemicals they will not dissolve in rain and so leach out of soil to accumulate in waterways to poison fish, vegetation, or other forms of wildlife; and they can be chosen to act in a predictable fashion against a pest without having any deleterious effects on predatory species selected for conservation. Viruses, by acting as transmissible genetic entities, certainly pose special problems – most of which arise from our inability fully to understand the relationship between viruses and the host cell they infect – so that, even if they may be our ultimate weapon in the war against pests, it would be premature to suggest that they are now ready to be used in this way. But microbe pesticides offer too much to remain in a limbo of scientific ignorance any longer.

Two fungi are of especial interest in the search for microbe control agents. They are *Beauvaria bassiana* and *Metarrhizium anisopliae*, and both have been used on a somewhat experimental basis since the beginning of the twentieth century. They are active insect pathogens, and their spores (which readily form) germinate on the body wall of the host, rather than having to be ingested with the food. The growing fungus colony penetrates the insect larva through a breathing pore, or spiracle. These are important advantages which make administration relatively easy. On the debit side there is the tendency for these fungi to spread only slowly through the tissues of the host, but this can be countered by the careful choice of the time of administration – and it is known that a high temperature and high atmospheric humidity help the fungi to develop. There are doubtless many other fungi that could be used to control pests by producing a specific infection – but one group, aptly known as the predacious fungi, actually ensnare their prey by the production of a kind of cell lassoo. These are soil species, and they can trap amoebae and other small creatures, particularly nematode worms. Nematodes are familiar to us as faecal roundworms seen in puppies and sometimes in children; and one microscopic nematode, *Trichinella spiralis*, is the species which produces trichinosis and is the origin of the Jewish custom of avoiding pork (in which this nematode is still sometimes found). To the gardener the nematodes are the minute eelworms which infect root crops.

The predacious fungi develop snares as they grow, each one about as large in diameter as the size of a soil nematode. As the prey, perhaps a rotifer or eelworm noses around and encounters one of these loops, the cell walls around it expand and tighten, holding it firm. In due course it is believed that

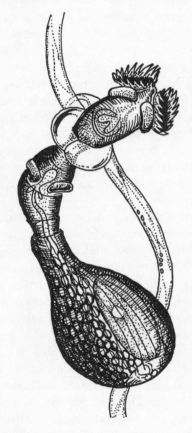

Fig 6: A rotifer, *Callidina*, lassoed in a loop produced by a predacious fungus. Rotifers and minute eel-worms are among the many-celled creatures which this type of fungus ensnares. The nutrients released as the prey breaks down are absorbed by the fungus, and predacious fungi in nature exert a control over many nematode worms which can damage crops.

its body contents are absorbed by the fungus which ensnared it. Some other fungi trap their prey on adhesive hyphae. The trapping of prey in such a deliberate manner is – to any biologist – an intriguing phenomenon. The possibility of getting microscopic pest-control agents to lassoo the enemy is one we have yet to explore.

Microbes are organisms whose bodies are not divided up into cells, so, as nematodes are many-celled, by definition they are not microbes. But before we pass them by it is worth noting that some plant diseases thought to be caused by nematode eelworms now seem to be the result of pathogenic bacteria they carry inside themselves. One such bacterium, *Achromobacter nematophilus*, has been observed inside a nematode and was seen to leave the worm through its anal pore after the 'carrier' worm had already entered the host plant's tissues. This is an interesting observation, for it suggests that we might be able to use nematodes to carry organisms inside a host. Indeed, there may be other ways in which we might utilize one species as a 'carrier' for another in just this way. So not only can we find microscopic species that will lassoo the enemy, but we can even find slightly larger organisms on which they might ride. The analogy between man's world and that of the microbe seems closer as we progress!

Other kinds of microbe may have still more possibilities for us to discover. Many protozoa infect particular species of pest, and could be used as pest-control agents, and a number of algae are known to be able to liberate by-products which exert powerful effects – though little is yet known about them. In an age when we will have to rely on pest control, and the side effects of many of our chemical products are known to be undesirable or even highly dangerous, microbes can usefully remind us that there are countless untried possibilities awaiting our attention. So here we are, feeling with no little desperation that our chemical insecticides are dangerous and that all our options have been taken up, when microbes offer almost unlimited possibilities which have still to be investigated.

The use of a microbe as a pest controller has something of the self-evident about it, but microbe power has implications for branches of industry that have nothing to do with biological mechanisms in the ordinary way. As an example, let us look at ways in which microbes could mine metals for us. One of the earlier sources of metals was a solution of the metal in natural waters. Thus in the 1500s copper was obtained from metal-carrying streams in Anglesey, North Wales; and in the 1700s it was similarly extracted from the waters of Rio Tinto, Spain. The salt present in both cases was principally copper sulphate, the blue crystals of which are familiar in elementary chemistry lessons. In nature we often find that the sulphate is present in streams because bacteria have been releasing the energy to reduce the ores to this soluble form. In the same way, 'fool's gold' – iron pyrites – can be changed from iron sulphide to iron sulphate by the action of these microbes, in which form it can be dissolved and leached out in solution. This valuable change in state can be of immense industrial significance, and it is important to realize that here too the energy for the effect is obtained from within the molecules of the substrate, and does not have to be supplied by man from outside sources. The microbe mainly responsible for these reactions in nature is *Thiobacillus ferro-oxidans*, an acid-producing bacterium well able to withstand strong solutions that would kill most forms of life outright.

At the present time, though little has been heard of the process, a dozen mines in North America carry out the release of copper from its ore through the activities of this microbe, and the concept could be readily adapted to the mining of low-grade ores throughout the world. It has applications for other metals as well, and the successful mining of uranium has been demonstrated by a similar pilot process. The microbe has the ability to enter a rocky shale or ore-bed and selectively release the desired element, thus removing the need to load and shift vast tonnages of wasted rock. The metals can be pumped, in solution, through pipes; which makes the whole mining operation potentially less labour-intensive and less expensive than it has been in the past.

Metals from other sources could be reclaimed in this way. There are valuable amounts of metals in refuse and spoil heaps and, though the concentrations are too low for orthodox mining and refining to be worth while, microbe mining could release metals from sludges and tips which for the moment have been abandoned. In terms of energy economics it is an attractive prospect. To carry out reclamation in the ordinary way would require the *energy* of men, the *energy* of machinery to dig the raw material, and *energy* to transport it. Large amounts of *energy* would be put into the system to separate the constituents. And finally, further *energy* would be required to bring about the processing of the ore before applying the *energy* of refinement. If the leaching system is used, then microbes eliminate the need for much of this energy demand. The copper pyrites (covellite) or the sulphide of another metal can release energy during oxidation. The microbe, once it has been put to work, can use the energy of this reaction to provide the power to continue the process as long as the reactants are present and the energy is derived from within the ore itself, a far more efficient procedure.

The liberation of trapped chemical energy underlies the alternative biochemical industries we might develop, now that petrochemical reserves are becoming increasingly expensive. Oil represents the stored solar energy of life forms that existed millions of years ago, and each tonne of oil burned is an irredeemable loss to us. I have long believed that in principle it is no better than burning banknotes for the sake of some temporary warmth, for oil contains so many valuable chemical raw materials that burning seems the last thing one should do with it. Furthermore, plants alive today can produce captive solar energy for us.

Our economy has been based for long enough on hydrocarbons (from oil); but in future we might instead opt for a carbohydrate economy (based on green plants). Thirty years ago we were all clothed in the products of current photosynthesis by green plants, either captured directly (as in cotton) or captured in grassland and then processed by mammalian metabolism to produce sheep's wool. But in either case, the solar energy that went into the make-up of our clothing was current, and renewable. Now half our clothes are synthetics. Not only are they produced by converting oil products, which are not renewable, but the energy of the conversion is obtained by burning fossil fuels. So at a time when our need for conservation is at a premium, we are consuming huge amounts of vital reserves and seriously depleting our store of fossil raw materials.

A carbohydrate economy would enable us to produce petrol substitutes and a whole range of plastic raw materials from the products of plants alive today. Some of them might be specially cultivated, others could simply be reused more efficiently for purposes more imaginative than we usually employ today; and many plant products that are wasted at the present time could be used as a raw material for energy production. It has been calculated, for example, that the amount of straw which, in the early 1970s, was disposed of by burning each year in Britain wasted 25 per cent more stored solar energy than the amount agriculture consumed each year from petroleum products. If we began to look for efficient methods of reutilizing such energy-rich wastes, we might find that large areas of industry and commerce might become self-powering, rather than energy-consumptive.

Solar energy is not the trivial source one might imagine from the scant interest that has been shown in it. The amount of energy reaching earth from the sun in three days exceeds all our known energy reserves. Neither is solar power as novel as we sometime. might imagine. Agriculture is an industrial system founded on stored solar energy, and its controlled release; and, before marvelling too much over an isolated solar-powered house or an unorthodox 'ecological home' with a greenhouse energy-collector, we would do well to reflect that man is a solar-powered device too; and functions more efficiently.

The photosynthetic process by which the energy of sunlight is trapped by a green plant is not highly efficient. Indeed for many plants it amounts to a rate measured in terms of parts per thousand. But the use of microscopic plants, such as algae, cuts out much of the wastage. Microbes exist bathed in their culture fluid, rather than relying on a complex system of waterways and gas-exchange structures like the stems and leaves of highly developed multicellular plants. Artificial lakes containing, say, sewage effluent could provide an ideal growth medium in which algae could proliferate. Alternatively, the intensive production of more conventional plants – such as pollarded trees or cereals – provides us with a starting point. One could even use seaweed.

But once we have an energy-rich carbohydrate product, there are innumerable avenues through which it can pass. Why should we merely burn it, releasing the energy in the form of heat, when we can more efficiently harness the energy to fuel the chemistry of life – which the microbe can then use to undertake the processing? If we have used hydrocarbons so easily in the past, why not utilize renewable carbohydrates for our future? Microbes that can degrade cellulose, the most complex and widespread carbohydrate of all, abound in nature (indeed it is reported that one American forces laboratory has collected 13,000 different organisms that can break down the cellulose molecule, releasing energy as they do so). To find species which could provide the basis of a cellulose economy should be simplicity itself.

Finally, if microbes can act as a new source of power in so many ways, what role could they play in making agriculture itself more efficient? It is basically true that green plants – through photosynthesis – produce elaborate carbohydrate molecules from the raw materials of carbon dioxide and water, nothing more. But many other chemical inputs are necessary too. Trace elements such as magnesium, sodium, and the rest, are important; and so are phosphates and sulphates. But by far the most important item in the inventory is nitrogen. It is this element which is a vital constituent of the proteins which are found in every living cell. Without nitrogen, a plant cannot grow.

Nitrogen is the commonest element in the atmosphere: four-fifths of the air is this gas. But in this form it cannot be utilized by a green plant. The nitrogen has to be combined with other elements, forming nitrates and ammonium compounds, before it can be incorporated into the structure of a green plant. For many years nitrates for soil dressing have been obtained from mines in countries such as Chile where beds of the natural material are found. In past decades there has been much investigation of artificial nitrates as a substitute for the mined material, the argument being that we were too reliant on geological reserves and that an artificial substitute would be infinitely preferable. Unfortunately, if we look into the mathematics of the process we can show that the production of 1 tonne of artificial nitrate consumes 5 tonnes of fossil fuel, so the dependence on geological reserves was not avoided, but transferred from one non-renewable source to another – and the demand was increased five-fold into the bargain. It is this form of short-sightedness which has underlain many industrial solutions we have sought in the past.

Even if artificial nitrates were energy-competitive there are other hazards attached to their use. In the soil they tend to produce nitric acid, which in turn has two effects: the acidity of the soil may increase, which can effect the health of plants grown in it, and secondly it reacts with limestone and dissolves it. Unless adequate lime is added to the treated soil the whole ecosystem can be disturbed. One possible answer to this which was popular in the earlier years of the twentieth century was the use of ammonia as fertilizer. As an alkali, it would presumably neutralize the effects of nitric acid in the soil, and it contains nitrogen in the form of NH_3 which is immediately available to green plants. But this idea was not wholly success-ful either. Ammonia is rapidly oxidized by microbes and chemical reactions in the soil, to form nitric acid. So in some respects the addition of a supposedly 'neutralizing' alkali had much the same effect as an administra-tion of acid – and the problem became even greater than before.

The hope of producing unlimited supplies of crops by the addition of extra nitrates or ammonia proved to be largely spurious. The widespread use of chemicals in the United States since the Second World War removed an enormous amount of limestone from the soil, a total that must now be many thousands of millions of tonnes. Yet all the time that we were using industrial methods to obtain a ready source of nitrogen for growing plants, there were many species of microbe already adapted to carrying out the same conver-sion without such drawbacks.

The best known are the nodule bacteria which grow on the roots of such plants as clover. If a clover plant is gently pulled out of the ground, the roots can be seen to be covered with small, wart-like excrescences. The roots look almost 'diseased'. These nodules contain colonies of bacteria – *Rhizobium* – which can take nitrogen from the air and combine it chemically to form the more complex compounds needed by the host plant. Bacteria of the *Rhizobium* type can exist well enough on their own, so that it is not necessary to argue that they need the protection of the plant's root for their survival. But their capacity to fix nitrogen directly from the air conveys considerable benefits to the leguminous plant on whose roots they occur. It is perhaps worth realizing that the meadow clover, in this way, is able to harness the microbes to undertake the manufacture of its own fertilizer, while man has yet to find a way of doing that.

Obviously we could save energy and raw chemicals if we could use microbes to fix nitrogen in an analogous manner, by harnessing *Rhizobium*. If more could be discovered about the mechanisms by which these bacteria enter roots then it might be possible to encourage them to form nodules on other plants which would open many avenues for the growth of crops in poor soils. A more adventurous approach is to isolate the genetic information inside *Rhizobium* which lays down the systems that fix nitrogen. If they could be added to the genetic codes naturally present in green plants we could even envisage genetic surgery which would make plants self-sufficient for nitrogen. On an experimental basis, some interest has been shown in all these possibilities. But I think there is a danger that we may become so obsessed by *Rhizobium* that we overlook the other kinds of microbe that can also fix nitrogen, and which may provide unheard-of alternatives. There may be many as yet undetected soil microbes which fix nitrogen for our crops.

Others may be at work far from land. It is one of the mysteries of the ecological network of the oceans that large areas of the sea contain far less fixed nitrogen than marine creatures need to grow. For example, shipworms feed on wood and obtain their energy by oxidizing cellulose. But this is only part of the story, for they grow at rates that demand high levels of fixed nitrogen and yet it is known that the wood on which they feed provides them with very low levels indeed – nitrogen levels in wood are almost always considerably less than one part per thousand.

If shipworms are grown in culture in can be shown that they fix nitrogen directly. The rates (around 1.5 microgrammes of nitrogen per milligramme of dry weight each hour) mean that cells of one of these worms could produce their own weight of new fixed nitrogen in less than two days, which is a highly efficient way of doing things. This explains how these creatures are able to grow rapidly when they take in only small amounts of nitrogen in the diet. But how can they do this? And why do terrestrial animals lack the ability? The theme of this book is that microbes often have the answer, and this example is no exception. Growing in the intestine of one marine worm has been found a bacterium which can indeed fix nitrogen. Most interesting of all, it feeds itself on cellulose and has been shown to be capable of liquefying cellulose in culture. This still unidentified organism explains how shipworms may obtain their supplies of fixed nitrogen – but it opens up the far wider possibility that it could be cultured and used to fix nitrogen for agriculture. The fact that an organism can exist on a diet of cellulose but can utilize nitrogen from the air to make new protein raises many hopes – and could help solve our fertilizer problem.

By comparison, it is now known that nitrogen-fixing algae are vital for the food supplies of coral atolls and reefs. The species concerned are 'lowly' blue-green algae, one of the smallest forms of microbe life. They grow in the form of a thin film covering parts of a reef, often too tenuous to be noticeable and too unattractive to warrant the same level of study as the dramatic and highly coloured corals themselves. But some of these algae, including certain species of *Hormothamnion* and *Calothrix*, can fix nitrogen at a rate that compares favourably with the highest performance of carefully controlled agricultural production. Here too there are some pointers to research that might be undertaken. These algae immediately explain how it is that the abundance of life on coral reefs is supported, in spite of the very low levels of dissolved food materials that are typical of the tropical seas in which corals are found. But as well as answering a long-standing biological puzzle, they also show how some microbes can produce as much new protein out of sea-water as mankind can obtain from intensively cultivated and well-fertilized farmland. The essential difference is that the algae use solar energy direct: there is no drain on resources, no power consumption, no pollutant chemical residue, and no waste.

Microbe power, then, is a serious business. It should not too readily invoke mental images of banner-waving microbes claiming 'germ liberation'. Microbes are well adapted to undertaking startlingly efficient chemical processes, and merely need to be recognized and channelled into new areas to become important sources of power for mankind. We have persisted for quite long enough in imagining that the demonstration of power, and the consumption of more and more energy, represent the technological equivalent of spear-brandishing. For a great many industrial processes the use of endogenous energy, already locked away inside the chemical structure of the raw materials, makes external demands on oil or other fuels unnecessary. Microbe metabolism can release the energy in such systems, and this form of self-fuelling reaction can allow a chemical process to take place without any external intervention beyond monitoring and regulating the rates of throughput and supply.

The principle of applying external energy to such systems – for all its instinctive appeal – is wasteful and, when we step back and look at it in perspective, ludicrous. The industrial answer to gathering a bunch of flowers would be to cut down and mechanically sort an area of woodland, so that the flowers emerged at the end of a conveyor belt. The biological answer would be to go out individually among the shrubs and vegetation to choose what you want, and fit them together as the fancy takes you. The industrial answer to collecting aluminium bottle-tops or cans would be to sift garbage and sort it automatically, while the biological alternative would be to have individual consumers drop them all together into a collecting box.

This is exactly the kind of new answer with which the microbe can provide us, by moving in among the molecules we wish to alter, and using their own internal life energy to power the process. Man's incentive may be aesthetic or intellectual gratification, or it may be a better pay-packet; but all that microbes require are raw materials and water. While we are facing an energy crisis and told on all sides that new sources of energy must be found (with the corollary that none seems to be available), a large proportion of the materials in our environment could be utilized to provide us with microbe power. Waste organic matter, wood, paper, grass-cuttings, metallic ores, agricultural refuse, and a limitless range of industrial by-products could become substrates supplying their own source of energy. The application of external energy to blast a chemical system into shape, producing smoke, ash, waste heat, and pollution, as our fossil fuels dwindle away, is absurd when microbes can move in among the reactants and carry out the task. Whether it is enzymes for the housewife or fertilizers for the farmer, the production of food additives or the mining of inaccessible metals, the microbes are all around us and waiting to be used.

Life *is* energy.

3 Food from the Microbe

Shortage of food is the greatest single problem facing the under-developed nations. It is more urgent than the need to limit population increase, since an immediately available contraceptive device would do far less to reduce suffering than an hypothetical never-ending supply of food. This is not to say that birth control, or public health measures, or anything else are unimportant; but it does remind us of our priorities. The provision of food is clearly Number One.

Microbes are the most protein-rich food one could find, and they also reproduce at rates that agriculture could never contemplate. The growth record is at the moment held by a species of *Pseudomonas*, and this bacterium can divide in two and double in weight in just nine minutes. Under ideal circumstances, then, 1 tonne of them could produce 1 tonne of new life every 9 minutes: how would a farmer react to a cow, or a pig, that could reproduce and grow to maturity in that space of time?

Let us quantify it more realistically. *Pseudomonas* is something of an exception, after all, and juggling with the figures for its rate of reproduction does not tell us what we need to know about the production of food for humans. Taking scaled-up versions of culture tanks now in existence we can calculate that we could provide more than enough food for the world's population in an area the size of an industrial estate. Or to express it more dramatically, one could have the population of our planet increase by one hundred times, say, to 400,000 million people, and still feed them from an area of Arctic waste land the size of London. This is the promise of microbe technology – it is the reality of cell food.

None of this suggests that we should throw all our conventional diets out of the window and eat nothing but micro-organisms. Eating orthodox foodstuffs from vegetables and fruit to meat and fish is here, as far as we can

41

calculate, to stay. The era of 'one white pill for breakfast, and a yellow tablet for lunch' is never likely to appear. Man enjoys his food. It has important social connotations and it is culturally significant to eat according to subconscious criteria learned from childhood. The pleasure of eating steak or fish and chips will always remain – even if prices put them into the luxury food class. The starving of the Third World countries, who are at present hard-pressed to obtain any appetizing food at all, will want to enjoy the flavours of chicken, pork, beef, and the rest in an age when hunger is banished. So we must not imagine that pills will take away the pleasure from eating: they are unlikely to be the sole diet even of spacemen.

In future we will have to recognize the wastage in producing many conventional foodstuffs. Cattle and sheep, for example, can locate and consume grass and produce a nourishing meat end-product. In effect this amounts to the processing of inedible grass and silage into a form acceptable to mankind. For the grazing of poor uplands there is no better way of obtaining food. But the process, for general purposes, is uneconomical. One-third of the energy from the grass that is eaten is wasted through respiration; one-sixth is lost in faeces and urine; and so on . . . so that only 4 per cent – less than one twentieth – is eventually available to the consumer. We cannot afford to waste 96 per cent of the stored solar energy in this way, at least in such a hungry world.

One answer is to look to the primary producers of protein from sunlight: green plants, in other words. Whereas beef production provides 30 kilogrammes of protein per hectare each year (protein for human consumption, that is), and instensive pig-houses provide 50 kilogrammes, the rates are far higher for plants. Cereal crops reach the level of 135 kilogrammes, and potatoes 325 kilogrammes. Figures from the United States suggest the cost of egg protein to be around 475 cents per kilogramme wholesale, and beef 350 cents; but the cost of protein from plant sources was around 60-80 cents. The sea could provide us with some unexpectedly high rates of protein production, too; fish farming is known to be able to give up to ten times as much protein per acre as beef production, and the total world catch of 7 million tonnes or so of fish from the oceans could be increased. To calculate how much this could amount to, we must consider the amount of energy reaching us from the sun which is wasted – 60 per cent is lost through cloud cover and scattering, and half of what is left is infrared wavelengths that cannot be utilized by green plants. Twenty per cent is lost by absorption by rocks and snow; and 10 per cent is reflected back from the surface of the sea.

Even so, the amounts of energy taken up by the oceans have allowed some calculations to suggest that we could catch 300 million tonnes of fish per year – so the energy intake of the oceans could in theory enable us to increase the amount of fish we take by forty times, without depleting their rates of reproduction. The starting point for the food supply of fish is marine microbes, and we might derive greater benefits by harvesting smaller creatures that are nearer to the primary sources of solar energy capture. Shrimps and plankton are replete with energy, and each time a larger creature devours a smaller one some 90 per cent of the energy of the food is lost. The netting of small sea creatures could provide nourishing food at levels amounting to thousands of millions of tonnes per year – and there is a precedent for the idea. Gauze nets are hung across tidal currents in parts of Indonesia, and the minute shrimps that are caught are dried in the sun to form a sea-food biscuit.

Other ways of locally boosting the sea's productive capacity include the deliberate run-off of phosphates and nitrogen – to fertilize the algae that capture sunlight – or the placing of heat sources at the bottom of the sea in areas where nutrients are trapped at depth. Small nuclear heaters would have the effect of raising food-rich waters so that algae could grow in the sunlit upper layers of the sea, and an extension of this principle could prodigiously reinforce the ocean's productivity. Then there is the possibility that algae could be cultured in sea-water to produce protein literally out of the air. The oceans are where life originated, and our own blood system is a portable sea in which each of our own cells is bathed – so the sea already has a fundamental niche in our development. As a long-term prospect for the provision of food its potentialities are incalculable.

But there are many short-term, more accessible answers to our problems. Microbes as producers of food are closer to us than we imagine. Meat-eating man likes to think that his beef or lamb derives directly from a diet of lush grass. But these animals cannot digest it. A cow fed on grass in the same way that we feed on vegetables would die of malnutrition. Cattle have accessory pouched stomach chambers, which are their own internal microbe reactors. Within these culture chambers grow large and complex communities of microbes, which break down the grass and silage on which the animal feeds and proliferate as a result. The cow's habit of regurgitating her food at intervals to chew the cud is her way of mixing the microbes with the food she has eaten. At the end of this processing, the cow digests the microbes that are produced and it is they which make up her real diet. The cow, then, is not herself digesting grass at all. She is feeding on microbes by the trillion.

The beef-steak so richly prized by Western man is made from the molecules of microbes which themselves processed the cow's intake of grass. More of these independent microbes are digested by the cow in a few seconds than the total number of cows that have ever existed. And this is not the only way in which man eats second-hand microbes: the vegetables that constitute a sizeable proportion of our diet obtain much of their nutriment from the decayed remains of the previous season's leaf-fall, which has been processed by the microbe population of the soil. Many of the plants on which we rely have a rhizosphere association involving their roots in a complex relationship with the microbes which provide them with raw materials for growth. Then there are the crops with nodules on their roots, in which bacteria fix atmospheric nitrogen as a source of protein-building materials. And even the green chlorophyll of a plant's leaves is contained in minute bodies that are probably ancient captive microbes put to work for the good of the plant – and serving quite coincidentally to provide us with food into the bargain.

One does not have to look far, then, to see how we already feed on the products of microbe processing. The dislike of microbes which we have inherited makes the idea of eating microbes seem inherently repugnant. Yet there is nothing very strange about it: we eat meat, and we enjoy vegetables. What are they if not masses – communities, if you prefer – of cells? If we are perfectly amenable to eating the small rounded cells that together make ox liver, why should we for an instant reject microbes that have been reared in fermentation vats instead? They are all cells – and their different origins do not necessarily imply a profoundly different product. Products such as Bovril, Oxo stock, MBT, Yeastrel, Herb-ox and Marmite are all famous proprietary protein-rich concentrates which are valuable additions to any diet. Yet the Bovril type is made from meat extracts, whilst the Yeastrel-like group are developed from microbes (from yeast cells, as the name implies). They make an interesting comparison, since in colour they are almost identical, in consistency they are similar, and even in taste they come into the same general category. Consumers might imagine that they had similar sources – yet one group comes from many-celled mammals and the other from the microbe.

Our tastes, our likes and dislikes, originate largely from habit and from learned criteria. We do not normally associate wholesomeness with microbes, and it is tempting to imagine that the dislike is fundamental to our nature, rather than resulting from a set of learned values. But many of our most appetizing foods would sound repugnant in the extreme if we described them in similarly emotive terms:

We snare a cold-blooded, slimy sea-creature and suffocate it slowly.

We split it open in its death-throes, spilling its entrails to the floor, and then we hang up the dripping, limp corpse.

The cold carcass is thrust into a choking fume-filled gas chamber where some poisonous tarry substances condense on to the drying, dead remains.

Then the disembowelled, half-mummified, stiff corpse is partly dried and charred before being torn to pieces and eaten by a man.

Yet this is no grotesque scene from an avante-garde film or a violent novel. I am talking about a kipper.

We are so quick to associate micobe protein, or cell food, with unattractiveness; but how would we react to that appetizing steak if beef was described as 'dead flesh hacked from the dismembered body of a loyal pet with its brain pulped or its throat cut'? Some commentators have gone so far as to insist that microbe protein produced from the use of oil as a growth medium will inevitably taste of oil – yet the protein extracted from yeasts is just protein, and it is no more sensible to warn that it will taste like the substrate in which it was produced than it is to claim that chicken will have to taste like hen-food, or lamb like scrub grass. Additionally important is the efficiency of cell food production. A single square mile of equipment could produce sufficient microbe protein to feed every man woman and child on every continent of our temperate, blue planet.

One of the most commercially promising avenues of progress in the move to develop cell food has been the growth of yeast cells in fuel oil. In cold climates, diesel oil tends to become too thick to flow properly, and this phenomenon occurs because waxes and fatty molecules in the oil become solid and begin to separate out. It has been a serious problem facing mechanics in cold countries for many years, and mechanical or chemical answers to the problem have never become economical – but microbes had already found out how to deal with the problem. The species concerned is a yeast, *Candida lipolytica*. It is related to an organism which exists in the human vagina, and its name means, literally, the fat-splitting yeast. *Candida* can feed on the unwanted waxy components of the oil, leaving the fuel purified and free-flowing. The yeast cells are separated out as a paste – and it was in the search for an application for it that *Candida* as a cattle-food was developed. Since the notion first appeared in the latter part of the 1950s, only partial success has resulted from translating the pilot experiment into a commercial process. British Petroleum built a plant in the mid-1960s at Grangemouth in Scotland, which yielded 4,000 tonnes per annum; and have since opened a unit at Lavera, France, whose initial production of 16,000 tonnes per annum has since been raised to 20,000 tonnes per annum with plans announced to increase that to 27,000 tonnes per annum. The first large-scale protein producing factory BP have announced has been built at Sarroch, Sardinia, in cooperation with the Italian ANIC company, and this was planned to produce around 100,000 tonnes per annum in the late 1970s. Two leading Japanese companies which announced proposals for giant plants subsequently revised their estimates downwards. Initial plans put forward by the Maruzen Oil Company gave Japan a pilot production rate of 6,000 tonnes per annum, and following this lead the Dainippon Ink and Chemical Company announced plans for a plant producing 120,000 tonnes per annum while the BP process licensed to the Kyowa Hakko Koggo Company looked for a time as though it was going to produce 100,000 tonnes per annum.

By the early 1970s one could have predicted that it was only a decade or so before Japan was likely to produce over 1 million tonnes of protein feedstock per annum – which amounted to almost half of the world's shortage. Yet (as keeps on happening in this field) interest waned. Critics of the process, ever alert to the problems of making microbe food acceptable, publicized the finding that there was a known carcinogen in the product – 3, 4-benzpyrene – and this, coupled with the public distaste for microbes, caused two companies to cancel their plans. The amounts of the impurity detected were less than one part in 2,000 million, and though it is laudable that concern should be shown over a finding such as this, the fact that tea, coffee, beef-steak, and toast can all contain very much higher levels that this was entirely overlooked. One would not want to encourage haste, especially in such an important field as the production of safe and nutritious food; but we would be in error in applying prohibitively rigorous standards to new foods unless we use them for our more traditional items of diet, too. To do justice to microbe protein, and to cell food generally, it would be wiser to limit the consumption of foods on the basis of known hazard, rather than making exceptions of foodstuffs solely because they were 'traditional'.

Japan is not the only nation to cut back on these experiments. BP's plant at Lavera was planned to produce 100,000 tonnes per annum by the mid-1970s, but plans for that expansion were also postponed. Little of the thinking behind this cautious attitude has its roots in fears for the safety of the product: so intensively is it monitored that microbe protein would be likely to set all-time records for purity and safety standards. The leading potential producers privately agree that it is public attitude which is the greatest obstacle to progress. Until we all revise our criteria, then interest will continue to be suppressed. That champion of industrial innovation, the United States, has hardly any plans for the development of cell food – and their largest yeast-producing plant of the type described above, aims to produce a mere 5,000 tonnes per annum by 1980.

But it is not only as a sideline of the oil industry that proteins can be produced. An Italian company, Liquichimica, has announced plans for a plant in the Calabria region of Italy which will produce several important chemical products (ranging from citric acid to lysine) from microbial fermentation and will also generate up to 200,000 tonnes per annum of cattle-feed made out of the microbes that undertake the biochemical work. Meanwhile, plans have been announced by Taiwan and Czechoslovakia for cattle-feed plants producing in excess of 80,000 tonnes per annum, and it is believed that a 200,000 tonnes per annum unit is planned for the Soviet Union.

We can see, then, that the energy in a molecule can be released in several different ways. The most wasteful is merely to burn it, which is what we usually do today with energy-rich molecules; the most useful is probably to utilize the same energy as fuel for the energies of life – energy which can be used by man. Oil is the example we have considered so far, since it is the best-known substrate of all. But any other energy-rich molecule could be a candidate for the approach. At present we are tapping natural gas supplies for use as fuel, but they could be diverted to microbe processing. In this way we could oxidize the gas biologically, and produce chemical by-products and food directly.

The traditional approach to this would be to burn the gas to raise steam, and from this generate electrical power which would provide the energy for conversion. The scheme could be outlined thus:

Microbe Power

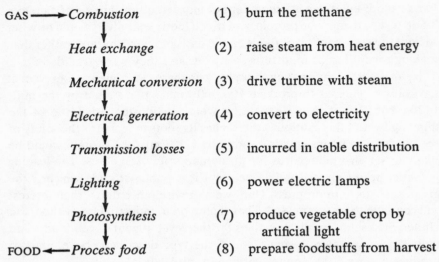

GAS ───▶ *Combustion*   (1)   burn the methane

         ↓

     Heat exchange   (2)   raise steam from heat energy

         ↓

     Mechanical conversion   (3)   drive turbine with steam

         ↓

     Electrical generation   (4)   convert to electricity

         ↓

     Transmission losses   (5)   incurred in cable distribution

         ↓

     Lighting   (6)   power electric lamps

         ↓

     Photosynthesis   (7)   produce vegetable crop by artificial light

         ↓

FOOD ◀─── *Process food*   (8)   prepare foodstuffs from harvest

It is tempting to put accurate figures on these stages, and derive an overall rating for the conversion of the gas into a food product. Some of these stages are efficient (e.g. electrical generation, which can rate around 90 per cent) while others – including photosynthesis, at around 3 per cent – are not. To quantify this scheme introduces too many unknown variable factors to make sense, but it is clearly a highly inefficient approach and the output of food would be a vanishingly small proportion of the fuel input. But it is through this kind of thinking that much of our industrial society has developed, and if the yields were small the traditional answer would doubtless be to increase the size of the plant – which is exactly the kind of thinking which has given us much of today's pollution burden.

By contrast, look at the alternative offered by microbe technology. In this alternative model we take the gas and allow the life chemistry of the microbe to oxidize it. The heat energy is converted directly to metabolic life energy, and food is produced directly from the gas:

GAS ───▶ *Metabolism*   (1)   microbes use methane as

FOOD ◀───   energy source

This is a more predictable system, and it is more realistic to put figures on it. In this case it is known that half the carbon 'fuel' is converted into cell substance by the microbe, and nine-tenths of the nitrogen added as a protein-builder is incorporated too. The system is therefore over 50 per cent efficient, and gives us food direct from natural gas. There are several

48

organisms known which can oxidize methane in this way, and their exploitation as a protein source could mean that a natural gas find could supply food,

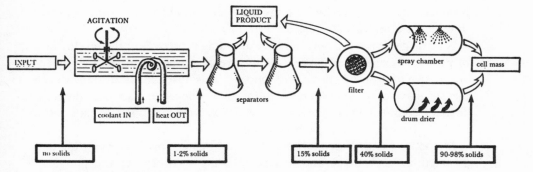

Diag 2: Cell technology: energy and food from waste. This simplified scheme illustrates how industrial or domestic wastes could be converted into valuable new products for society. Pollutants, for instance, can become raw materials and food without needing external energy for the conversion.

rather than an easily expendable form of heat, for a developing nation. Certainly we should have investigated these possibilities earlier and, with a different climate of opinion, we doubtless would have done so. The first accounts of methane-oxidizing microbes date from the 1800s, when German microbiologists noted their existence.

Suggestions that methane-oxidizing organisms might be a useful source of protein were first made prior to the First World War, but there has been little progress since. The natural gas boom in the North Sea during the 1960s encouraged some leading British companies to investigate ways of harnessing the methane molecule directly, and attempts were made to culture microbes that could oxidize the gas. But methane is an energy-rich molecule, and the amounts of energy produced by the growth of the microbes led to over-heating. It was felt by the engineers involved that cooling costs would make the process impracticable: others might have argued that the waste heat produced was itself a valuable by-product.

But there are several organisms known to oxidize methane, including *Bacillus methanicus, Methylococcus capsulatus*, and the *Methanomonas* organism. Some of them are capable of enduring high temperatures, so cooling is not so necessary. A high temperature (around 45°C) would introduce an additional benefit, too, for the majority of other organisms could not endure it, and this would make it far easier to ensure that unwanted microbes did not proliferate in the unit. In many ways, the growth

49

of bacteria on methane is the perfect way to use the gas effectively since we obtain a range of complex chemical substances, instead of burning the methane straight back to carbon dioxide and water. These are the ultimate products, in any event – but how much better it would be to use the range of intermediate products on the way, (p. 98).

Natural gas is not the only unlikely sounding substrate for the production of protein. Microbe technology can derive energy from other fuels as an alternative to the crude process of burning. This poses an interesting principle, since oxidation is the key; and as this is the underlying factor in common between combustion and life we can predict that any combustible material is likely to be a microbe growth medium.

Methanol (methyl alcohol) could be used as a growth medium for the production of proteins, and this helps overcome some of the practical difficulties that have been attached to the proposals to grow microbes on methane. Methanol is already partly oxidized: if a single atom of oxygen is added to the methane molecule, methanol is formed:

$$H - \overset{\overset{\text{H}}{|}}{\underset{\underset{\text{H}}{|}}{C}} - H + O \longrightarrow H - \overset{\overset{\text{H}}{|}}{\underset{\underset{\text{H}}{|}}{C}} - OH$$

methane + oxygen   gives   *methanol*

Its further breakdown to water and carbon dioxide produces less surplus heat than the oxidation of methane. Of course, the energy levels being lower, we produce less protein in the end. The methane-oxidizing bacteria we have referred to earlier can produce yields of 4 or 5 grammes of protein for each litre of culture fluid, whereas the microbes that oxidize methanol – *Hyphomicrobium* and *Pseudomonas extorquens* being the best known – grow to give yields of around 3 grammes. *Hyphomicrobium* also has a slower growth rate than many microbes, dividing every 14 hours or so. On the other hand, methanol is readily soluble in water, which makes it easy to produce a controllable growth medium. Methane is not, and has to be circulated round the microbes throughout their growth.

But the various factors in favour of one idea or another are difficult to assess. This is new technology, indeed it is largely untried, and we are limited in our understanding of just what is involved. There may yet be other forms of microbe which could be used for the purpose: at least one strain that has been considered was almost accidentally isolated from the gardens around the laboratory. There are other fuels we can consider. In Czechoslovakia, Switzerland, and the United States there are pilot plants in the planning stage which may show how to use ethanol (the alcohol we drink) as a growth medium. Paraffin and fuel oil could also be utilized. There are a dozen or so microbes already known which can utilize these hydrocarbons as a growth medium, and doubtless many more awaiting discovery. Certainly the use of petroleum to produce proteins and other by-products is far better than burning it away.

The culmination of this approach is the production of foodstuffs from wastes, and almost any energy-containing refuse is suitable for the purpose, whether sewage, waste paper, or unusable refuse from agriculture. Many materials regarded as pollutants or effluent today could become growth media for food-forming microbes in the future – a dramatic reversal of roles. Waste sugary solutions are common enough in the food-processing industries, and they exemplify the kind of effluent that we could convert. Using 100 grammes of hydrocarbon, such as oil or methane, we could theoretically produce roughly 100 grammes of protein. But the amount of oxygen consumed in the process amounts to 200 grammes.

Now – if we had started with 100 grammes of carbohydrate instead – we would produce only 50 grammes of protein, but the oxygen requirement would be lower still – a mere 35 grammes. So we are producing half as much protein, true; but we need to inject only one-third as much oxygen. The technology involved in bringing oxygen into contact with growing microbes involves a good deal of extraneous stirring, pumping, and filtration equipment; and the consumption of five times as much oxygen liberates five times as much heat energy. This also has to be disposed of and, as we have seen, this is an additional problem for the engineer.

Costings show that the present state of technology makes carbohydrate wastes preferable to hydrocarbons. The cheapest substrate would be fuel oil, which would cost around 2p for each kilogramme of cell mass produced. But the extra cost entailed in additional cooling and oxygenation would be as high as 4.7p. Using molasses as a substrate the cost per kilogramme of product is higher – over 5p. But in this case there is no extra investment needed for these ancillary items, and so the final product is actually cheaper.

At present the carbohydrate-containing wastes (of which molasses are an example) have excellent commercial potential for the production of food, and there are many sources of the waste which would be ideal for conversion. A pilot plant has been tried out by Rank Hovis McDougall in Britain, which uses a starch waste obtained as a by-product from the processing of bread and bread products, yams, potatoes, and other sources. Ammonium salts are added as a source of fixed nitrogen, and the mix is inoculated with the *Penicillium* fungus. The growth which results contains 50 per cent protein or more. A similar experiment has been carried out at the Louisiana State University in which the pulped residues from sugar cane and rice straw are fermented by the rod-shaped bacterium *Cellulomonas* which can degrade the cell-wall structures of the waste, *and* then digest the carbohydrates released.

One kilogramme of protein is obtained from every 5 kilogrammes of the waste, in this experiment. At the GE Research and Development Laboratories at Schenectady, New York, it has been shown that the growth of cellulose-degrading organisms can be increased by the addition of sewage sludge. These trace elements and nitrogenous wastes in sewage are ordinarily considered to be a pollutant, but in this experiment they became a valuable food-production aid. Clearly a system which produced cattle-feed from sewage and cellulose waste would be of immense economic, humanitarian, and ecological importance. The treatment of sewage to provide us with a protein product is not novel, in the sense that this kind of composted material has been used since time immemorial as a fertilizer. The liberation of methane by microbes that degrade sewage has been exploited in many pilot experiments, and sewage farms often use the methane they produce as a by-product to power their machinery or to light their offices. But here too we are crudely harnessing the energies of microbes that could, in suitable circumstances, produce more valuable results.

There is no requirement for this form of microbe technology to be on the grand scale, either. The 100,000 tonnes-per-annum plants which have been proposed would cost £30 million to set up, which is a form of investment it would be difficult to justify in an under-developed, Third World country. But – as a compost heap reminds us, and as the chicken-shit engine testifies – microbes can work on a less grandiose scale, and in apparatus that is exceedingly cheap to erect. In this way wastes could be recovered in the form of cattle-feed, without the establishment of elaborate factories. In 1975 a small-scale plant was being erected in the South American farming community of Belize, to recycle wastes from the fruit-growing activities in the area. The costings – based on the figures for 1972 – showed that more than 2000 tonnes of pig and poultry feed were purchased by the community at a cost of around £140 per tonne. The total expenditure for that year was £313,632. Yet the citrus orchards in the area were themselves producing well over 2,000 tonnes per annum of plant waste, the majority of which was dumped and caused pollution problems of its own. Scientists from the sugar trading company Tate & Lyle Ltd calculated that the waste could be used as a growth substrate for the production of cattle feed by microbes, the product costing about £90 per tonne. It was to test this in practice that the Belize pilot plant was planned. As this book goes to press the system has yet to be given a trial run but what remains to me the greatest puzzle of all is how such a low-cost system has had to wait until now to be tested at all.

There are many other wastes that could be adapted as raw materials for microbe technology, and for the production of foodstuffs. Paper and cotton wastes could be used, and the degradable portion of household refuse is an obvious alternative. Other residues that could be considered include grape pulp, discarded leaves and trimmings from vegetable-processing factories, and even grass-cuttings and harvest left-overs . . . and so, too, could inedible vegetation, such as the jungle undergrowth, be used. One enterprising manufacturer of tea and coffee-bags in the United States is investigating the possibility that spent tea-leaves and coffee-grounds could be used as a microbe substrate and converted economically into cattle-cake.

What we have to realize is that these organic waste materials have been elaborated by solar energy into an infinitely complicated range of chemical compounds. The microbe enables us to harness these, to convert them into end-products demanded by the pressures of civilized living, to end with protein foodstuffs – and to do all this through the chemical energy contained in the molecules themselves.

Yet the amount of attention paid to the principle remains scant. To take just one example, in 1971 a spokesman of the Rockefeller Institute, which is sponsoring research into how microbes degrade waste, was reported as saying: 'It is even possible that human food will one day be manufactured out of waste materials.' It has been 'possible' for a considerable time: and the 'one day' is already with us.

Many chemical effluents from heavy industry could also act as a growth-medium for microbes. We may be able to find simple and inexpensive ways of converting steel-works waste water or phenol-containing effluent from coal mines into food for animals, or even man. Microbes can metabolize many such wastes, and microbiologists at the Enserwerke Laboratories in Switzerland have already demonstrated one example. They took the waste from the manufacture of nylon (the principle component of the effluent being cyclohexane) and found that, after treatment, it could be utilized as a growth medium for organisms which could be formed into a high-protein concentrate. Food from effluents – what better way could there be of solving two pressing problems at the same time?

So far we have looked at ways in which the microbe can release the energy stored in chemical structures, but this is not the only possible solution to the problem of food production by microbes. We could grow the microbes that capture solar energy and store it – the algae. Solar energy falling onto a bed of algal cells would allow them to reproduce rapidly with the production of a food material out of nothing more elaborate than air and water, plus the admixture of a little sewage or industrial effluent to supplement the nutritive value of the brew. The output of the system would be cell food and oxygen.

No such utopian solution exists without practical difficulties, however. The process of photosynthesis poses special problems, by far the greatest of which is the need for the culture chamber to be extensive, and shallow. Deep fermentation vats are inapplicable to a situation which requires sunlight to reach every living cell. In addition, the areas where food from sunlight would be the greatest benefit are the tropical zones where water itself is often at a premium. Some form of closed container could be necessary for the control of evaporation losses, but this would add to the expense and further reduce the levels of sunlight admitted to the culture. Experiments in Japan and in Czechoslovakia have been disappointing, because many of the difficulties were not foreseen. But in principle the idea is simplicity itself, and there are many species of algae that might be used.

The species that grow in filaments can be harvested by simply raking them out of the water, and a blue-green alga, *Spirulina maxima*, grows in twisted threads like a corkscrew, which knit together to form a tight compact mass which is easy to collect. An algal system would be ideal for development as a source of food supplements for astronauts. The algae would absorb man's waste carbon dioxide, his recycled water, and some excreted waste chemicals, providing foodstuffs and fresh supplies of oxygen in return, the system being powered by sunlight.

Although this source of food is perhaps the most futuristic of all, it is one with an ancient origin. Algae have been harvested for centuries (and perhaps for thousands of years) by the native population that live on the shores of Lake Tchad, Niger. The lake is fringed with firm, dark soil and the water level rises during the winter season, only to retreat again each spring as the water table shifts. The lake water is rich in fish, and in the summer season large mats of algae appear – the warm, shallow water and the high levels of sunlight are ideal for their growth. They are dragged out of the water at harvest-time and left to dry in the sun to form solid cakes. When required for food, the dried substance is pulped with water and stewed to form a nourishing soup. This form of solar-powered microbe food has become a staple item of the region's diet. Little wonder people say there's nothing new under the sun.

And could we not utilize other sources of energy? Protein could be produced from man-made energy sources, such as hydroelectric power or the nuclear reactor. Electrical generation has an important drawback: it is that the continuous production of energy results in overproduction at off-peak periods, while the generation capacity may be overloaded when demand is highest. It is inevitable that, if plant is available to reach the highest levels of load normally expected, then it will be under-used for the bulk of the time. Tidal storage systems, or the pumping of water into high reservoirs, has been used in some sites to consume power and use it to move water against gravity. Release of flow when demand for electricity rises again allows a proportion of the energy to be regained.

But it is possible to use the off-peak electricity to produce hydrogen for microbes to metabolize. Electrolysis of sea water would provide hydrogen and oxygen (the water, H_2O, is split into its 2H and O constituents) and the bacterium *Hydrogenomonas* is able to derive energy for its own life processes through the reverse reaction. In this way the gases revert to water, and the microbe-protein is collected as a valuable end-product, either for use direct, or perhaps in turn as a substrate for further processing. In this way we could even convert surplus electricity into foodstuffs, and if we look further we could include such diverse sources of energy as tidal generation, wave power, and even the wind in the list of food sources we have yet to exploit.

Whether cell food becomes immediately available for human consumption, or is used instead as a feedstock for animals, depends as much on the nutritive value of the product as anything else. It is convenient to confine the discussion to protein production, but protein is not the only component that matters; and there are many different form of protein, some of which are inadequate as dietary supplements. The correct balance of component amino acids is one factor we have to consider; and high levels of nucleic acid – such as are found in bacterial protein – are not suitable for human consumption. The United Nations Advisory Group suggests that 2 grammes per day is the ceiling for nucleic acid consumption, and bacterial protein contains up to one-fifth nucleic acid. Purine, a chemical which is a feature of many microbe foods, is metabolized differently by man and other mammals. In most animals the end-product of purine metabolism is allantoin, which readily dissolves in the body fluids. In man, by contrast, purine forms uric acid and this forms insoluble crystals causing gout. So one could not assume that any form of protein was automatically suitable as a staple diet for mankind.

On the other hand, this has always been the case. A varied and balanced diet is well known to be necessary for health, and we would become unhealthy if we ate nothing but bacon, or butter, or eggs. The observation that one particular kind of microbe protein is inadequate as a whole diet is no counter-indication for its use, since all foods have to be viewed as part of a complex dietary regimen. Microbe foodstuffs contain amino acids lacking from some cereals and they can rectify many deficiencies in traditional diets.

The rate at which cell food could be produced is difficult to comprehend, and even harder to relate to our everyday notions of proliferation. Some plant species exist for hundreds of years before doubling their number. The human species has a generation time measured in decades: yet some microbes can do this in a matter of minutes. In terms of a single cell on a microscope slide this is not particularly significant at first sight, but what we have to remember is this: in dividing from one cell into two of the same size, in 20 minutes or so, the microbe is doubling its weight. It has to duplicate every feature of the parent cell, every molecular complex, every organelle, every particle of the original structure. It is more dramatic when we realize that to a single cell this is equivalent to, say, a sheep or pig producing a full-blown copy of itself every 20 minutes, and for both of them to keep doing so.

Starting with a single cell, then, we would have 2 at the end of 20 minutes and 8 at the end of 1 hour. But if we had started with a tonne of microbes, we would have *8 tonnes* at the end of that time. Theoretically, to produce the almost 3 million tonnes of protein food needed to rectify the global annual deficit, all you would need to do would be to take 300,000 tonnes of microbes, wait for a little over an hour, and you would find the extra 2,700,000 tonnes produced as if by magic. Indeed we could just as easily start with a single microbe, feed it up for a week or so, and theoretically end up with food outweighing the entire solar system.

But that is theory. In practice such dramatic schemes are limited by lower rates of microbe growth, the need to provide suitable substrates, to circulate nutriments, to regulate temperature and to harvest the product. When we realize that a microbe can reproduce every 9 or 10 minutes, we can perform the most staggering mathematical athletics – but we must not lose sight of such practical considerations. On the other hand, it is possible greatly to increase the metabolic throughput of a microbe which normally divides every few hours by supplying it artificially with an ideal environment. The boosting of a normally limited nutritional requirement can increase the performance of a cell, and to some extent we can always obtain better results from a microbe in culture than would ever apply in nature. Microbes which have evolved for the rigours of a competitive existence can produce biochemical changes at very much amplified rates when a technologist removes constraints.

Using up to date intensive methods of meat stock rearing one can double the weight of beef animals in about 3 months. A young pig can double its weight in 4 weeks, a chicken (or a field of vegetable crops) in half that time. When microbes are grown in today's commercial fermentation chambers, they can double their weight at a maintained rate *within 2 hours*. These are data obtained from accurately timed trials; not hypothesis. A small factory producing microbe food could turn out 1 tonne of protein per day from each of, say, ten fermentation vats. To attain the same level of production of protein from conventional sources you would have to produce pigs for slaughter at the rate of 100 animals per day; to carry out the same conversion with grain as the starting point would demand the use of up to 4,000 hectares of farmland. The microbe factory would produce high-quality foodstuffs from an area one twenty thousandth of the size. Indeed a centre producing enough to feed the entire world could be erected in a few square kilometres of the Arctic waste lands, where the supplies of ice-cold sea water would dissipate the heat generated by the life processes of microbes working incessantly for man.

Before we can see this matter of food production in its proper perspective – and before we can understand the role of cell food – there are some fundamental considerations that have to be taken into account. How many people are hungry at the moment? How crowded is our planet? And where are the prime target areas?

In the under-developed nations that comprise some 70 per cent of the world's population, average diets provide something like 6 per cent less nutriment than is considered necessary for a healthy existence. The developed nations of the West show consumption figures for calories that rate 20 per cent above the limits necessary for health, and average body weight is significantly greater than those known to be associated with longevity. But we have to retain a balanced view, and many of the claims about birth rate and living standards are misleading. Japan, West Germany, and Britain all have a population density greater than 200 persons per square kilometre. The United States, so often cited as a land of crowded communities and chronic over-population, has a density ten times less (and only twice the levels of some countries with large areas that are largely unpopulated, such as the Soviet Union and Brazil). In terms of population size, the 1968 figures show that China, India, the Soviet Union, and the United

States – in that order – are the four largest nations in the world. But those that show the fastest rate of increase (all in excess of 2 per cent in the period from 1963 to 1968) are India, Brazil, Indonesia, and Pakistan. Definitive relationships have been claimed to exist between population size and rate of population growth; between overcrowded cities and population density; between degree of development and population increase. But none of them apply to the real situation which faces us – and the simple observation that North America is one of the world's more sparsely populated areas, yet has more problems from urban overcrowding than almost anywhere else, makes nonsense of many popular beliefs about the human community.

It is repeatedly argued, for instance, that peasant peoples *have* to exist in areas prone to natural disasters, while the more affluent head off to find safer havens. Thus the well-off congregate in the secure regions of the world far from the monsoons of the Bengal lowlands and the peasant huts in earthquake zones. This cannot be reconciled with the facts. Look at those huge populations in California that live straddling the San Andreas fault, all in line for a monstrously disastrous earthquake if the predictions of geology and common sense prove right. Or look again at the regular toll to life, limb, and property caused by the hurricanes along the Florida coastline; yet this area is a uniquely powerful magnet for the affluent capitalist in search of the 'better things in life' for which he craves.

People live in a given area not just because they have to, or because they lack the wherewithal to move somewhere else, but because they feel settled and secure there. This is why populations drift towards towns, cities, and the super-urban masses. There may be pollution, but man feels safety in numbers – a psychological instinct that the statistics refute. In 1920, 12 per cent of the population in the world's richest nations lived in cities housing more than 500,000 people; by 1980 the figure will be 28 per cent. In the underdeveloped nations the figure in 1920 was 1 per cent; it will be 13 per cent by 1980. This is the problem for the future – the question of security, of belonging, of morale. Though the proponents of orthodox demography sometimes seem to predict a situation in which the world is entirely overcrowded, like people rubbing shoulders in a busy street, the 1970 figure of population density shows that the world's human community could increase almost tenfold before the people were as crowded as they are in Britain – and anyone who has looked at the map of the British Isles, or who flies over its territory, must have been struck by the large proportion of sparsely populated land which makes up the bulk of the country. This is not to say that population levels are unimportant, far from it; but it does emphasize that it is our ability to deal with ourselves and to feed, clothe, and provide for people that matters, not the immediate limitation of how many of us there are. In addition (as the population figures from Great Britain are likely to reaffirm), increases in the cost of living may cut the growth of population in developed countries to near replacement level, perhaps even less.

At present rates of world population increase, cell-food and microbe technology could allow us a century or more before population levels began to pose problems, and could reveal thousands of new ways in which we could use replaceable energy. Long before then we will have readily available and safe forms of on-demand contraceptives for both sexes, produced by microbes in industry. In physical terms, then, the world seems to be facing insurmountable problems. But the microbe can solve them: it is now a question of attitude.

For a century we have feared the microbe, and now we have to come to term with microbes as allies, as indispensable co-workers. There is no reason why we should consider microbes distasteful, apart from the bias of the way we have been taught to look at them. I doubt whether you could argue that, because some bulls have killed men, beef itself should be distasteful; but this is the kind of attitude we have applied to microbes. Indeed, it is perfectly possible that foods made from cooked and seasoned typhoid bacteria, or any other pathogen, might be perfectly delightful to eat. I have eaten cooked cultures of bacteria prepared in my own laboratory and was not unduly deterred by the origin of the material. But that does not mean I would gladly want to eat an anthrax pâté for dinner, or tuberculosis soup for my lunch: to be frank, I think that I would find the notion distasteful in the extreme. But we must be aware of the learned responses that give rise to this attitude. Like the rogue bull, the capacity of an organism to kill has no great bearing on its taste or on its palatability.

Yet when occasional snippets of information have been released by food companies, mentioning some futuristic opinion that a yeast might be used to provide us with a protein supplement, the newspapers are often quick to decry it as 'fungus food'. 'Who wants to eat mould?' is the reaction: 'We are faced with the threat of synthetic and artificial foodstuffs.' No one is deterred from consuming beer or wine because they derive from microbes; or from enjoying cheese or yoghurt because much of them is composed of bacterial cells. And of course, there is nothing 'unnatural' about microbes: as we have seen, the kinds we could develop into food have been in existence for thousands of times longer than mankind, and all our present-day food derives directly or indirectly from microbes in any case.

Microbe Power

The argument that 'processed' foods are unnatural clashes conspicuously with the claims of wholesome food fanatics that a product such as stone-milled, wheat-germ bread is 'natural'. The origins of the technology may have been lost in antiquity, but when you go into the sophisticated background of such everyday things as a cake, a cup of tea or coffee, salami, or a pair of kippers, then what is so remarkable is that such complicated forms of artificial processing could have been derived in relatively primitive times. Or were the people alive then merely showing the kind of inventiveness that goes hand in hand with an open-minded approach to food?

The reason, I am sure, is that people living in those times treated food in whatever way gave them the best results, without any of the squeamishness which we know about. If their primitive techniques used the activities of microbes – even though their very existence was undreamed of – they were adopted if they worked in practice, an attitude of simple pragmatism. Our own distaste is relatively recent in historical terms, and seems to be coincident with the propaganda surrounding the germ theory of disease. I imagine that meat that is 'high' (a connoisseur's delicacy) has been far less popular since bacteria were presented to us as enemies. For centuries it was prized, just as cheeses are today. Though we have gained much in terms of public health since then, we have lost out on the uses to which the microbe – as an ally – could have been put. In the earliest phases of mankind's existence, before even the simplest ways of processing foodstuffs became available, the quest for food must have been mankind's prime motivation. Plants exist in contact with their supplies of water and food, and it is the need to find the essentials for existence which has given higher animals the ability to move, and the diverse sense organs used to detect them.

Man's weakness and vulnerability would soon have caused his extinction, had he not the brain to outwit and to trap his prey. We have been taught that primitive man obtained his meat by hunting, indeed, the phrase 'man the hunter' is virtually a household expression. Yet I doubt whether this was the case. How would man hunt, without claws or fang? How would he survive with so little protection from an adversary? How could he catch his prey with such limited running ability (for man is surely one of the slowest mammals of all)? Man can never have been a primitive hunter. No, he needed his brain, his cunning ways, for a different purpose – to *outwit* other animals. It was he who could wheedle his prey over a cliff, or into a pit or a cave. Man's upright gait and his uniquely manoeuvrable hands were his means of building traps and weapons with which to overcome stronger and mightier species. 'Man the hunter' was really *man the trapper*.

The feeling of hunger, to primitive man, was a signal we can now recognize as ensuring he met a physiological need – he would instinctively want to feed until the hunger signal abated, and then he would stop. Civilized man, however, prostitutes this instinct. If he lives in one of the developed (over-developed, some might say) nations then the stimulus for him to eat is the time of day, or the odour of appetizing cooking. Then he eats, and eats; and eats. He only stops when he is 'feeling full'. But this is a warning; it means he has eaten too much. Man in his sophisticated environment of artificially enhanced flavours and predetermined ideas on being chubby, well-built, and the need to eat a 'hearty' meal, is misapplying these innate desires. In an earlier era we needed hunger to drive us to eat: but now that we have identified the flavours that attract us and know how to intensify them hunger becomes a drive to over-eat. This is the reason why Western man is dying through obesity, heart trouble, sclerotic arteries, and the rest. Fatty substances, rich in energy, are still liberated into our bloodstream at times of stress, but whereas in primitive times they would have been burned up in fight or flight they can now accumulate in our blood vessels as we 'grit our teeth and bear it'. So the criteria by which Stone-Age man selected his options no longer fit our life-style. Unless this aspect of the human machine is taken into account our problems in relating food to civilization are likely to intensify, and not disappear.

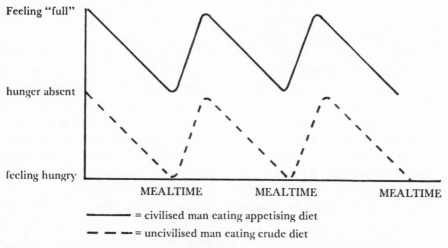

Diag 3: How food technology and the development of appetizing foodstuffs affect eating habits. Civilisation brings with it a tendency to be perpetually overfed, in contrast to the conditions under which we evolved.

Microbe Power

One much-neglected aspect of research into the provision of food is this relationship between evolution when a diet is primitive and plain, and existence in a modern societal system geared to the values of gratification and the technology of flavour enhancement. On the other hand, if we have moved so easily from a primitive diet to a processed one, and from that to a largely artificial intake of bland, bleached bread and convenience foods, then the incorporation of cell food into the daily menu should be, by comparison, a small step to take.

Of course, cell food, if taken to mean nothing more than microbe protein, will not provide a complete diet, and even the balancing of constituents to ensure the optimum intake of fats, vitamins, and amino acids does not obviate the need for the more basic, physical constituents of food. Already the Western diet of refined, soft foodstuffs contains too little roughage for the intestine to function correctly. It is usual to find food passing from mouth to anus in as little as thirty-five hours when people eat a normally bulked diet, containing adequate fibrous cell-wall material. But in civilized and sedentary people eating a 'pap' diet these transit times may easily exceed a week. Though it would be undesirable to invoke an epidemic of bowel-consciousness, which is one of the less desirable obsessions with which we afflict ourselves, it is certainly time more attention was paid to the fact that we seem perfectly content to allow food wastes (some of which may be potentially toxic or carcinogenic) to remain in the absorptive tract for such protracted periods. The addition of roughage to our diets, whether obtained from crops or from fibre-producing cell cultures, could itself increase the health of any 'civilized' community.

So the development of cell food will have to take its place in a broader reconsideration of mankind's nutritive requirements, and a fundamental reassessment of the dangers inherent in many of our accepted values. For too long now we have ignored the biological background to our instincts and our habit patterns, and in the future it should be possible to look again at man's background, so that his life-style takes more account of what is truly 'natural' for him. The resulting pattern of life, based on realistic options chosen for pragmatic criteria, could be better fitted to our needs in practice and can open our eyes to a diet that is more normal, more natural, and better balanced than our diet has ever been. We may all end up fitter, better fed and healthier by far.

And that includes the starving millions.

4 The Microbe against Pollution

Living in an industrial society has presented us with a serious and widely publicized pollution problem. The wastes of industry and the discarded refuse of the typical Western-style family have begun to accumulate faster than we can dispose of them. But we have already seen how the growth of microbes in industrial wastes provides many possibilities for the production of proteins and other edible materials, and the upgrading of simple molecules (like methane) by microbes – which could themselves be converted to other chemical substances – broadens the field further. This brings us to one of the most timely possibilities of all: that the microbe could convert those pollutants into harmless substances, and could help cut the effluent burden to a fraction of its present level.

Microbes are already doing this in nature. Many of the proposals for effluent control disregard microbe power – indeed, the folly of the 'non-microbe' approach is best summed up by a novel idea put forward in 1973, that in future our rivers (at present often little more than open sewers) should become 'clear-water conduits' instead. The error of this approach is that if rivers were made totally clean and all the water was filtered and processed to become entirely pure, they would soon end up in a worse state than they are in at present. The purity of a clear, sparkling river is due to the activities of its microbe populations. Were the rivers to be purified and the microbes in them systematically removed, the first algal cell that alighted in the water would rapidly turn the river into a thick soup-like brew, since there would be no other microbes to keep matters under control. A glance through a microscope at the bottom of a stream will reveal the microbes that purify the water. Without them, the water would not 'sparkle' for long. As children we were brought up never to drink such water 'in case there are microbes in it'. On the contrary, one can drink it because it is so pure, and it is kept in that

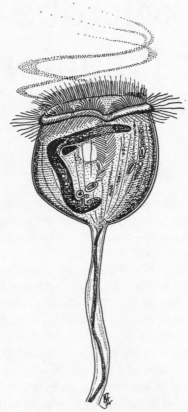

Fig 7: The 'bell animalcule', *Vorticella*, plays a vital role in keeping fresh water wholesome. Its rows of beating cilia set up a current of water, from which the microbe extracts bacteria and other small forms of life to use as food. Each cell acts as a selective and highly efficient filter, and even the most advanced technological device could not imitate them.

state by microbes of a myriad varieties.

If we are going to clean up our rivers, the first ally we need to recruit is this hidden work-force. Microbes, given encouragement, can do most of the reclamation work for us. To start with, it is not good enough to speak of a lake or river as being 'poisoned' or 'dead'. Though they have been over-looked in the main, there are many microbes that can exist on inorganic chemicals, and they are at work in the polluted areas, reducing the amounts of effluent. Britain's River Trent and North America's Lake Erie are both famous examples of extreme pollution. They are the textbook example of a 'poisoned river' and a 'dead lake'. But how do they look if we bring the microbe into the argument?

The Trent is the third longest river in Britain, though it tops the list in terms of pollution. Shakespeare wrote of it as the 'smug and silver Trent' but today it is so filthy that none of the five million people who live in the Trent Basin drink water from it. At the present time its state is so bad that it is considered unpurifiable. The origin of the problem lies in the amount of industrial development which sprung up in the area. The Trent flows from the Potteries to pass Stafford and to collect the effluent of Birmingham and the River Tame; it is joined by the Derwent and the Dove from industrial Derbyshire; then it flows past Nottingham and on to Scunthorpe and the Humber. The list sounds like a catalogue of prime sites from the Industrial Revolution. One-third of Britain's coal is mined in the Trent area, and it is said that one-third of all British power stations stand on its banks. The use of the river for cooling purposes in the power plants raises the temperature of the water as much as 5°C, and this causes one-third of a million cubic metres of water to be lost through extra evaporation in a single day, which serves only to concentrate the pollutants still further. Meanwhile the huge new housing estates that have sprung up everywhere overload outdated sewage systems so that, in heavy rain, raw sewage pours into the river as the sewers overflow.

There are many ways of combating the problem: filtration, screening, chemical neutralization, flocculation to bring down the suspended solid particles, purification of the effluents by the factory producing them by industrial processing, and so on. Once a river is in a highly polluted state, however, it is difficult to know where to start – and no one assumed the task could be undertaken without the expenditure of large amounts of capital, and the use of large amounts of energy, which in turn creates new problems of its own. But are there not some interesting alternatives, which allow the natural processes of microbe life to take a hand? Indeed there are. Many centuries ago it was common for the sewage and waste water from small communities to be run into a lake or pond in which fish were present. The water used to remain pretty clean in spite of the burden that was regularly present. In some areas of the world exactly the same principle is in use to this day and some of the water from the Rhine in Germany is run into lakes and allowed to 'regenerate' in the same way. But what happens to all the pollutants? They have to go somewhere, and chemical oxidation cannot account for more than a small portion of the total.

What is happening is that microbes tackle the effluent and effectively recycle it. They were doing so in the earlier stages of the Trent's history, even after industrialization. As recently as 1887, when industrialization of the area was advanced, 3,000 salmon were taken from that river so it is apparent that the levels of pollution at that time were compatible with the river's microbial cleansing systems. If only they had the time to act, before being grossly overloaded with newly released material to handle, microbes could clean up the river for us even now. Some recent trials have shown how simple it would be to arrange.

In one of them, an artificial lake was dug by the river at Elford; in another experiment a disused gravel pit at Lea Marston was flooded with river water which had previously been run through three large settling tanks. Most of the coarse sediment settled out in the tanks, and the still cloudy water was run to a depth of 2 metres into the gravel pit. There was no artificial stirring or agitation of any kind: the water was left to its own devices – and the microbes got to work. Bacteria that could feed on phenols and other poisonous chemicals appeared, and in the purer water on a sunny afternoon the green algae began to multiply. Sometimes the algae produced blooms at the surface, looking like what you might call 'stagnant water,' but in fact a teeming community of innocuous, industrious microbes.

Within months, small crustaceans such as the water-fleas *Daphnia* and *Cyclops* appeared, grazing on the algae. Beetles, fly larvae, and other aquatic creatures began to increase in numbers. Within a year a stable community had evolved. Fungi and bacteria were disposing of the chemicals in the water, protozoa were eating them and in turn feeding the water-fleas; trace elements left in the water were taken in by algae which synthesized protein for the fish and smaller creatures; and the water issuing from the outfall of the experimental lagoon was, more often than not, clear and clean. A throughput of 180 cubic metres per hour could be obtained from the 1.1 hectare system, and it was soon found that – even on its worst days – the near-finished water could easily be treated to provide a potable supply.

The data which were obtained from the lagoon showed that amounts of nickel in the water were largely unaltered, though toxic elements such as copper and zinc were cut to one-third of their initial levels. Cyanide and phenol vanished altogether. Pesticides were present in the river at levels below one part per billion, and they did not significantly decrease. But the bacteria from sewage, the ubiquitous *E coli* which is used as an indicator of the presence of faeces, were cut by 96 per cent. The sludge that collected at the bottom of the lake consisted of microbe protein, mineral particles, and metallic industrial effluent. A typical summer analysis, as part of the Trent Research Programme, showed the following levels:

TABLE 2: ANALYSIS OF POLLUTION SLUDGE FROM THE RIVER TRENT

Metal	*Concentration* (in parts per million)
cadmium	47
nickel	608
lead	617
chromium	1,340
copper	1,815
zinc	4,520
iron	51,500

The principle result of this astonishingly simple and inexpensive process, in which nothing is done but to allow the river the chance for its natural microbe populations time to deal with civilization's discarded wastes, is

Fig 8: The sun microbe, *Actinophrys sol*, rotates slowly as it floats in still water, where it feeds on small algae. With its straight, translucent radiating projections it looks like nothing more than a child's stereotyped picture of the sun. It extends rounded, transparent amoeboid processes when it traps its food, and if the prey is large enough several of the individual microbes join forces to attack it together, separating when digestion is complete. *Actinophrys* is an important natural means of controlling water pollution.

water of a purity that is close to that of a natural river. It is ideal for use in industry, and the predicted future demand for water from the Trent makes this an most important consideration. The predicted demand for the year 2000 is in excess of 1,500 million litres per day compared with 180 million litres per day in 1975. If water can be recycled more efficiently then this huge demand on an already overloaded river may well prove to be an over-estimation. It is known that there were 17 litres of river water to dilute each litre of effluent back in 1912; in 1965 the ratio had fallen, so that there were only 5 litres to each litre of effluent; and the trend allows one to predict that by the year 2000 the load will double yet again, if unchecked.

It is a predicted future that the microbe could dramatically alter. If experimental lakes can provide all that is necessary for river water to be reclaimed after being grossly polluted, then it should be possible to extend this principle to waste land, old quarries, and gravel pits in many areas. Indeed, might not the answer to many polluted rivers be to construct these microbe lagoons along the banks? What about regulating reservoirs at intervals? There are many permutations – but the one conclusion that is inescapable is that expansive and costly plants would be superfluous and, if the microbe was given an opportunity to get to work, our rivers could recover within a decade.

What about that other product of microbe purification, the sludge that collects at the lake bottom? The immediate suggestions put forward were, of course, dumping and incineration. But there may be many ways in which the metals in the deposit could be reclaimed, and microbe mining could be applied to the task of separating out the metallic residues from the inert components of the sludge (see p. 34). Whatever the eventual answer proves to be, readily available microbe populations could overcome even this pollution problem.

Let us turn to Lake Erie, almost 25,000 square kilometres of polluted water. Surely, microbes could be given a chance to reclaim it in the same way? One of the most interesting lessons from the lake's history is how efficiently its indigenous microbe community controlled its pollution in the past. The suggestion that the new upsurge of industrial development around the lake was damaging its ecological systems was first made at the time of the First World War. Surveys did not confirm the suspicion, however; indeed as late as 1928 an ecological examination of the Lake concluded: 'Nowhere in the Lake is objectionable pollution of any kind found.'

At the time the conclusion seems to have caused some surprise. Now we can see why it was that the lake remained pollution-free for so long. Its teeming microbe population was coping adequately with the effluent load, utilizing it as a food material, and keeping the water wholesome. Since 1960 conditions in the lake have rapidly deteriorated. At first sight there is ample justification for the books that state that Lake Erie has a 'fatal disease', and the broadcasts that describe it as a 'dead lake'. But the lake is not really as dead as these gloomy reports suggest. Microbes are still there, in force; and much of the pollution poured ceaselessly into the lake is still broken down by microbes to form harmless materials. Algae still survive, indeed in summer they occasionally form huge algal blooms that spread across the lake, colouring its surface green. At the other end of the ecological scale, fish still abound. They are not the same species as were most widespread at the turn of the twentieth century, but are kinds of fish that are more tolerant of polluted conditions. None the less, the lake still produces the same tonnage of fish each year as it did around 1900. There is still plenty of life there, and, given encouragement, it could surely recover its ecological life-support system. We say that a body of water is eutrophic when there are more chemicals in solution than can be accomodated by the cyclical changes that naturally occur. But the ultimate fate of the effluent degraded by microbes is that toxic compounds are changed into organic substances and, in this form, they *can* enter the ecological network once more. Indirectly, every tonne of fish taken from the water represents reprocessed effluent!

There are many indications that the activities of microbes could assist us in overcoming our growing pollution problem. Their role in maintaining the purity of natural bodies of water has been demonstrated many times, whether through direct experimentation or through observations on ecological systems in rivers, lakes, and the sea. The scope of their activities, and the quantities of materials that microbes handle, are still open to debate. But if the Trent has been poisoned by the exploitive myopia of British industry, and Lake Erie has an ecological structure that has been distorted and overstrained by the capitalist materialism of America, then the Mediterranean is a miraculous survivor of the wholesale outpouring of effluent by the French, the Italians, the Spanish, the Albanians – and many more. Though the Mediterranean is not quite a land-locked sea, it takes a century for the water in it to be complete exchanged through the Straits of Gibraltar. The loss of water by evaporation is reflected in an increase in the levels of salinity in the sea, which ranges from 36.2 per cent at the western extremity to 40 per cent off the coast of Israel. Everyone knows of the 'filth' of the Mediterranean, and of the large amounts of effluent that are daily discharged into it.

Microbe Power

But what we ought to realize is how remarkable it is that the bulk of that sea remains crystal clear. The shores of the Mediterranean are internationally famous as holiday resorts, because of the warm clarity of the sea. Forget the occasional muddy outfall from a river, where turbulence mixes up everything from the sea-bed into a turbid mass; or the bays where enormous industrial loading has caused the water to turn murky. Most of the Mediterranean is crystal clear, and the majority of its bays are clean and rightly famous for their pale sandy beaches and glass-like transparent sea. There is no better testimony to the ability of organisms in nature to control pollution. As well as the sea, and the beaches, we ought also to appreciate the incalculably large and complex microbe population which is incessantly at work in the water, consuming the pollutants, metabolizing the sewage, degrading toxic molecules to harmless substances, and keeping the bulk of the sea clean; unbidden, unappreciated, with no capital outlay – and against all the odds.

There is, of course, one form of pollution which is widespread, troublesome in the extreme, and persistent. It is – oil. The very term covers a wide spectrum of substances, ranging from volatile petrols to heavy mineral oil; and including secretions of the skin and the energy-rich plant oils. The oil from oil-wells, known commercially as 'crude', consists of a range of molecules ranging from the dissolved gases which are present in small amounts, and which – though hydrocarbons – are too light to be oils at all; right up to the long-chain molecules which have a consistency like pitch or tar. Microbes can break down the smaller molecules (we have already considered the way they can oxidize methane and other forms of hydrocarbon fuel – see p. 50), but as the size of the molecules increases, and we move from light oils through progressively more viscous and denser materials, it becomes more difficult for microbes to degrade them. Spilled crude, then, tends to lose its lighter fractions through evaporation and through microbe attack, while the components that remain become steadily more sticky and objectionable. These are the most troublesome residues, and they are the least amenable to microbe control.

Even so, the ability of microbes to attack and degrade oil has certainly been under-estimated. Much of this is understandable, since the oils are not abundant in nature – the oil-beds are usually quite out of reach of living microbes that might evolve towards their use as a food material – and, since oil is so immiscible with water, it is difficult for microbes to get at it at all. But given time, even the most recalcitrant oil spillages seem to be mopped up by microbe activity. The best example must be the after effects of the disastrous

70

spillage of oil when the Torrey Canyon was wrecked off the British coast in 1967 and the beaches of Cornwall were left thickly coated with a treacly mass of oil. Many of the news services tried to assess how long the damage would last, and the bulk of specialist opinion suggested that the oil would be there for decades to come. One of the official spokesmen who issued a statement at that time even went so far as to put a time limit on the persistence: 'as much as forty years'. At the time it certainly looked grim, but microbes usually seem to undertake more than one realizes, and to do it faster than one could have imagined; so, when asked about it, I was bound to reply that I believed the sea's ecological system might destroy all the oil within one or two seasons. Those comments were based on the respect one grows to have for microbes and their efficiency, but even so the rapidity with which the oil disappeared was greater than any of the estimates. By the end of the *same* year the beaches were mostly back to normal!

Are there more effective ways of putting microbes to work on oil spillages, than merely relying on them to seek out the target and get to work in their own time? This question becomes particularly important when we are faced with a disastrous spillage on a vital area of coast: tourism, or the maintenance of bird colonies, are factors which can make the prevention of damage into a top-priority matter. The orthodox answer to spillage is the washing of the oil with detergent in an attempt to disperse it into the sea-water. This has many unsatisfactory results, not least of which is the wholesale destruction of wide areas of marine life by the action of the detergent on living membranes, and it has only a limited application to huge masses of floating oil or thick beach-side deposits, where we encounter the practical problems of getting the detergent into contact with deeper layers of the oil. One answer to this is the use of sand or ash literally to sink the oil.

Sand or ash is first treated with silicone (pulverized fuel ash from power stations, PFA, is often used in practice as it is widely available as 'a waste-product in search of uses') and this has two important effects. It makes the particles strongly water-repellent, so that they float on the water instead of sinking. In addition, the silicone is strongly attracted to oil which means that a particle coming into contact with floating oil will at once adhere to it. When enough of the treated particles have done so, the oil becomes heavier and sinks to the bottom – carrying with it a considerable amount of air bubbles, trapped among the 'dry' silicone-treated grains.

So we have the oil, in combination with some wholly inert ash or sand, lying on the sea-bed in association with plenty of trapped air. This is an ideal situation for microbes which might degrade the oil. What we could do is encircle a floating slick with silicone-treated PFA inoculated with some of the microbes which are known to degrade hydrocarbons, so that the entire spillage would disappear to the bottom of the sea where it was digested by microbe activity. Cost is no limitation, since silicone can make particles non-wettable even if the coating layer is only a matter of the molecules thick. There are drawbacks, since the temperature of the sea makes the degradation process inevitably slow; and, of course, in certain areas a layer of oil on the sea-bed would cause problems of its own. But this is one way of getting rid of an oil-slick quickly and relatively innocuously, and there is no surer means of protecting important areas of coastline, and encouraging microscopic oil-removal workers to undertake the eventual destruction of the pollutant.

A great deal of oxygen is needed for the oxidation of oil, and under the sea the availability of oxygen becomes a limiting factor. It has been calculated that the dissolved oxygen content of perhaps 400,000 litres of sea-water is needed to oxidize one litre of crude. Yet in the face of the practical limitations, microbe populations continue to control much of the oil spillage we incur in our hasty quest for irreplaceable energy. Without microbes our beaches would long since have vanished under a permanent layer of glutinous tar; as it is, this ability to slowly destroy oil – and to increase in numbers as they do so – returns the pollution to the environment. As the microbes are eventually eaten by larger organisms, the crude on which they fed (and from which they obtained their energy) is converted into raw materials for the use of marine life generally.

Pollution of the farmland on which we depend for our food is a major problem too, but man-made chemicals in the soil can also be controlled by the ever-enterprising microbe. DDT has long been famous as a persistent insecticide, yet it has been shown that the common bacterium *Aerobacter aerogenes* can partially break down the DDT molecules, so it is possible that one day the last traces of that insecticide lingering in nature may have been disposed of and recycled by microbes. There are signs that this is happening already. It was generally predicted that, because of accumulation, DDT levels in wildlife would continue to rise for a period after the substance was withdrawn from use. In many areas, however, it has been found that the levels in fish-eating birds started to drop almost immediately after the misuse of DDT was curtailed. It seems to me almost certain that microbes in nature are already disposing of this surplus in ways we have yet to discover. Most of the sprays and pesticides developed by our chemical indistries are eventually made harmless by microbe activity. Ever the powerful herbicide

2,4-D can be degraded by several organisms. The fact that 2,4-D can be broken down in the soil was first noted in 1949; in 1950 the microbe *Bacterium globiforme* was named as the organism responsible for the phenomenon. By 1952, two more microbes had been added to the list (*Corynebacterium* and *Flavobacterium*); and since then several more bacteria, together with the mould *Aspergillus*, have been found to have the same proclivity. If we search through the literature we can find other examples of a microbe that has been found to remove a pesticide which man has added to the environment. *Corynebacterium, Clostridium*, and *Lipomyces* have been found to degrade Paraquat (the weedkiller for which there is no known antidote), and there are half a dozen genera than can destroy such herbicides as IPC and Dalapon, among others.

What new approach could we adopt, then, to bring pesticide research more in line with the activities of the earth's microbe populations? To begin with we could try to make the molecular structure of the product as suitable as possible for the microbe to remove from the environment once the desired control effect had been obtained. There are some clues already emerging from the study of these materials and their biological properties. For example, a branched molecular configuration seems to be harder for a microbe to attack, than a simple straight chain. The presence of chlorine side-branches hinders microbial degradation as well. Though these are not hard-and-fast rules, they do point the way to further research which could make redundant pesticides *easily* degradable. In a properly managed system, chemical pesticides would probably be rarely needed; but if they were selected with the appetite of the microbe in mind, long-term persistence would cease to pose problems. As it is, we owe most of the disappearance of pesticides from our environment to our magnificently industrious microbe neighbours.

The next step is to use microbes as pesticide producers, of course; and there are many ways in which microbes themselves could be set to control pest species (see chapter 2). The principle of cell culture applied to higher plants might provide further sources. The insecticide pyrethrum is a plant product, for instance, and there is no reason why we should not culture the pyrethrum-producing cells to give us a readily available source of the compound – and perhaps some experimentally induced mutations could provide alternative compounds for insects that became resistant in the future. As a final line of approach, we might discover other health-promoting microbes which serve to confer pest-resistance on crops or farm animals – but that is a topic to which I will return later.

Microbe Power

We also have to understand that the widely taught principle, 'natural compounds are easily degraded in nature, but synthetic ones are not', is misleading. We have considered some man-made pesticides which microbes can degrade easily enough, and it is worth noting that there are some 'natural' compounds which do not degrade in that way. Chemicals such as the humulones, tannins, exine, and other biologically produced molecules can resist degradation. Some of them have been found intact and unaltered after thousands of years – so 'naturalness' is no guarantee of degradability or anything else.

Some areas of land, however, are polluted by the poisons left over from an earlier age of industry. These are the industrial deserts that remind us of the old metalworks of a century or two ago in Britain. It was in those times, in the cradle of industrialization, when legislative controls of industry were few and long-term pollution hazards undreamed of, that effluents and the risk of occupational poisoning were far more dangerous than they are today. In the industrial North and Midlands of England, as in the Lower Swansea Valley in Wales, lie barren areas of ground. A description like that conjures up a picture of weeds, scrubby grassland, neglected stunted overgrowth; but in some of these areas the ground is truly empty. No scrappy tuft of turf survives in the bleak, burned earth, a legacy of the polluters of that earlier age. There are over 45,000 hectares of derelict land in Britain, all in need of cosmetic treatment at the very least, and much of it barren and poisonous to this day.

Some of the land is unable to support wildlife because of poisoning by lead, iron, copper, nickel, silver, arsenic, or cobalt. Other areas are barren because the sulphurous fumes of the late seventeenth and early eighteenth centuries killed the natural vegetation, and subsequent erosion carried away the organic content of the soil. In some of the more favoured areas you can find straggling colonies of the bent-grass *Agrostis tenuis* or the wavy hair-grass *Deschampsia flexuosa* braving the gravelly ground, and a few hardy species of moss may be found; but that is all. Zinc can be toxic to plant life at a concentration of 5 parts per million; in the Lower Swansea Valley it lies in the spoil heaps in concentrations up to 130 parts per million.

Life is not, however, entirely extinct in those areas. Microbes that can tolerate toxic chemicals are present, and they show survival is possible even in the most polluted regions. Quite recently some specially resistant strains of higher plants have been developed specifically for use in these poisoned soils, but although they certainly show unusual abilities to tolerate toxic metals, difficulties arise when they are used to recolonize previously barren areas. What the ground lacks is a fully fledged microbe population. Some experiments in the Lower Swansea Valley showed that even a dressing of top-soil was insufficient to restart the system. One could hardly imagine a more 'normal' base in which to establish growing plants, but the dissolved metals in the polluted ground disturbed the well-balanced microbe populations in the new top-soil – and without microbes, the soil seemed to be an inefficient substrate.

What alternative could one find? Tests soon showed that there were two kinds of dressing that gave far better results: sewage, and wet domestic refuse. In both cases there were such large microbe populations that – even with the disruptive effects produced by the pollution in the ground – enough survived to re-establish a community. Though we still know little about the exact identity of the various types involved and the way they influence each other, the return of visible signs of life to the poisoned ground was an important victory for microbe power acting on its own in a way science cannot imitate.

Many areas of Britain, having use glass milk bottles for years, are now turning to plastic or waxed-paper containers instead. One can understand why – it has been said that 5,000 million milk bottles were 'disappearing' in Britain each year, amounting to roughly 1 million tonnes of glass. This single example illustrates the extent of the refuse problem. But some items are already recycled; not as efficiently as one would like, perhaps, but are converted into useful raw materials once again. In Britain each year one million old cars are recycled, 500,000 tonnes of bones reemerge as gelatine, glue, or fertilizer; half of Britain's steel comes from scrap, saving over £200 million in imports, along with 35 per cent of the aluminium, the 26 per cent of zinc, 65 per cent of the lead, and over 40 per cent of copper. Waste paper is widely utilized (in spite of a general belief that it is not), and it has been said that each year £60 million is saved on imported wood pulp by recycling nearly one-third of Britain's paper waste – a record for any nation, save the Japanese, who claim 50 per cent recovery. Even rubber, that most recalcitrant waste, is being processed – 20 million old tyres and inner tubes are converted into 200,000 tonnes of elastic bands, rubber-soled shoes, Wellington boots, and carpet underlay so the dawn of 'recycling awareness' is already apparent.

Composting, however, is rarely used. In Britain roughly 1 per cent of all refuse is treated in this way, whilst the technique has lost out altogether in the United States. It is in America that the 'throw-away package' has been exploited to the limit. What with disposable cooking pans and baking tins, boil-in bags and spray-on cream, salad dressing and the rest, there seems no end to the waste. Solid refuse amounts to roughly 5,000 million tonnes per annum, including 250 million tonnes purely from domestic garbage. Americans throw away some 30 million tonnes or more or paper, some 5 million tonnes of plastic, 100 million car tyres, and enough empty cans and bottles to give a dozen or two to every human being alive today. The technology of recycling has proved to be costly, and so dumping of refuse is still widely used in the United States.

Yet how much we could learn by studying the way that microbes recycle wastes. They have been doing it for more than half the world's life-span. It is becoming quite popular to print on the fly-leaf of new books: 'This book has been printed on recycled paper'. In a broader sense, *all* paper is bound to be 'recycled'. It may be burned: in which case the stored energy is spent in a brief period of combustion, and the constituents are dispersed into the atmosphere or the refuse tip. It may be composted, when the oxidation occurs more slowly; but in any event the principle products (water and carbon dioxide) will eventually become part of the next generation of raw materials for life. But unless paper is sealed in a vacuum chamber and kept in the dark in perpetuity, it will inevitably be recycled!

The simplest form of microbial recycling is the compost heap, in which large populations of microbes are encouraged to work through organic wastes, converting them into a nutriment-rich concentrate ideal for addition to garden soil. Composting in a larger scale could be adapted for the disposal and recycling of town and city refuse on a large scale, and has some important advantages over the usual methods of handling:

● In incineration, as in composting, the organic matter is broken down by oxidation. Whereas an incinerator needs external fuel supplies to keep it going, the microbes in the composting process derive the energy they require from the refuse itself.

● Combustible material has to be near-dry before it will burn properly. Wet refuse is a considerable drain on the efficiency of an incinerator, and large amounts of energy may be consumed in evaporating the contained water before burning can take place.

● For microbes, some moisture is vital and so wet refuse is easily reduced to compost. The cost of drying is eliminated, indeed drying would be harmful to the success of the process.

● The gases from an incinerator's exhaust flue, which may contain toxic vapours, may be swept across large areas of countryside and can constitute a serious pollution risk. The principle emissions from a composting plant are water vapour and carbon dioxide; smaller amounts of some other gases are released, just as they are from microbe activity in the soil.

● In combustion, smoke and dust particles may be produced, whereas all these components are returned to the soil after composting. Residual or non-burned carbon is largely eliminated as each organic molecule is systematically oxidized by microbes to produce raw materials for other living things. Combustion does not give us the energy-rich end product we can derive from microbial composting.

● The compost process imitates nature by returning to the earth materials that originated there. The process is self-fuelling and reliable, and imitates the recycling of leaves and organic wastes in the ground in addition to being a means of retaining captive solar energy.

One example of the way in which the refuse of an entire township can be composted is in East Germany. Troubled by increasing levels of river pollution in the rivers of this flat and low-lying country, the GDR government enacted legislation which banned effluent discharge into waterways, and began to clean up many municipal dumping sites. Now, all the refuse from the 300,000 inhabitants of Dresden is composted. Larger items of rubbish are screened out, and magnetic separators are used to extract the ferrous scrap. The rest is mechanically shredded into a fine granular mass which is fermented by aerobic (oxygen-using) organisms. The end result is a useful soil conditioner – but if piles of the compost are left to mature in the open air a wetter, more manure-like fertilizer is obtained. Experiments at a horticulture centre near Berlin have shown that the quality of crops grown in the compost does not vary significantly from season to season, in spite of the fact that some periodic changes do occur (owing to such factors as the burning of brown lignite as a fuel during the winter months, instead of coal). For some crops the high-lignite compost seems to be an improvement, as the table reveals:

TABLE 3: RATE OF INCREASE IN CROP PRODUCTION FROM COMPOSTED REFUSE

Compost	*Crop*	
	Cabbage	Spinach
Winter (high lignite ash)	24%	9%
Summer (low lignite ash)	11%	4%

At present, Dresden has been the only GDR town operating such a system, but it was planned to extend the project to include the industrial towns of Leipzig and Rostock on the Baltic, together with Potsdam, in 1976-7.

Though composting is more efficient than incineration, it still relies on the recycling of energy on a long-term basis. The energy may not emerge in a single flamboyant instant, as it does in the incinerator, but it is wasted so far as the needs of society are concerned. Are there no better ways in which the microbe could extract the energy from refuse in a more accessible form? There are, of course; and the development of microbe technology could allow us to obtain a proportion of our fuel from the stored chemical energy in refuse. One way of making the process easy to develop is to rely on conventional industrial treatment for part of the conversion, as in an hypothetical scheme worked out recently at the Open University. Our industrial set-up has always been so resistant to novelty – whether it is the steam engine or spray steel-making – that microbe technology may continue to be passed over if it seems too revolutionary. So a processing scheme partly based on conventional methods might stand a better chance of being accepted. It involves the treatment of complex carbohydrates partly to break them down into sugar-like substances. This is done by chemical treatment. In the resulting liquid, microbes can be cultured to convert the product to alcohol. The sale of the alcohol is envisaged to make the process self-sufficient, even profitable.

And this is an important consideration since the disposal of refuse is becoming increasingly expensive. To take a recent example, a plant commissioned by the Greater London Council at Edmonton in 1970 cost £7,640,000 to built, and in a working day converted 1,300 tonnes of refuse into 260 tonnes of clinker, 147 tonnes of metal scrap, and 75 tonnes of ash. The refuse was reduced to around one-third of its original weight, but it cost £3 or more per tonne to carry out the conversion, which is hardly an inexpensive way of doing it. Some of the chemical energy trapped in the structure of the refuse contributed to the maintenance of the burning, and there were obvious end-products: clinker for hard-core, metal for recycling, and granular material for land-filling.

A similar principle underlies the American CPU-400 equipment, which burns solid wastes in a fluidised bed incinerator. From its daily capacity of 400 tonnes of refuse it is reported to produce up to 15,000 kilowatts of electrical power, and much of the heat output that might otherwise be wasted is used to dry refuse or sewage sludges.

But a cardinal principle of the microbe revolution is that the chemical energy can be utilized, not by burning, but by supporting the life processes of the microbes that can undertake a more efficient conversion. Instead of converting waste to compost, it is equally feasible to use it to grow organisms which could produce energy-rich substances, such as alcohol. If domestic waste paper and similar cellulose-rich materials are treated with hot, dilute acid solutions the complex carbohydrates tend to change to simpler compounds such as sugars. These are an ideal substrate for the large-scale culture of yeasts.

This alternative approach to the handling of waste begins with the grinding of refuse into small fragments. Heavy and inert materials, such as glass and stone, can be settled out; magnetic separators can select the ferrous scrap; and the lighter card and paper wastes can be separated from plastic refuse and pulped. The cleaned pulp would pass through a continuous-process device which would mix it with 0.4 per cent sulphuric acid at a temperature of 230° for 1.2 minutes. At the end of that time the acid would be neutralized by calcium carbonate, and the pressurized liquids would be cooled by passing them through a heat exchanger cooled by the incoming pulp. In this way the untreated inputs would be partly heated, which serves to recycle some of the heat energy. The hydrolysed pulp is then fermented by yeasts, and the alcohol is separated by distillation in the normal way. Only the first stage of the process is a purely 'chemical' treatment, then; the real work of conversion is undertaken by microbes. An alternative approach would be to use enzymes produced by microbes such as *Botrytis* to split the waste carbohydrate complexes into sugars.

If we assume that the refuse contains 40-50 per cent paper and card, which is a representative figure in parts of Britain, then the sale of the alcohol end-product could be calculated to represent a profit of 38p per tonne of incoming refuse. Extra paper waste increases the cost-effectiveness of the plant, so that if the percentage of paper and card rose to 60 per cent (the figure already typical of the United States) the profit would rise to £2.72 per tonne of refuse. The capital cost of the Edmonton incineration unit was stated to be £5,730 per tonne of refuse handled each day, and the alcohol-producing scheme costs up at a similar rate: £5,870. But the fermentation plant would produce an energy-rich fuel from the refuse, which enables us to have access to the stored chemical energy. The current demand for alcohol runs at around 500,000 tonnes per annum in Britain, and 2½ million tonnes per annum in the United States, much of it being used as a chemical raw material for subsequent conversion. Both figures are rising steadily, so the microbe plant could provide a much-needed raw material from our domestic refuse. We may not have thought of it in this way before, but refuse is replete with solar energy – and at the moment that is something we cannot afford to ignore.

Plastics are different. Microbes do not take well to a diet of plastic, and it has become well known that plastics are non-degradable. This fact has been seen as a serious problem for the future, when plastics will be more widely used and – because of their non-degradable nature – more difficult than ever to dispose of. But the true state of affairs is not, I think, as bleak as all that. There are some microbes which affect plastics, many of them by damaging the plasticizers which go into making the finished article. Many plastics are rigid or brittle, and plasticizers are the additives which give them the flexible and soft quality so often needed. Microbes that attack the plasticizer tend to make plastics more easily broken up, and more likely to perish. Even in the years of the Second World War some 150 plasticizers were known, and half of those were shown to be capable of supporting the growth of fungi. The discovery that microbes attack plasticizers is not a new one, then.

It has recently been found that some bacteria can multiply on the surface of plastic articles, and plastic sheeting can sometimes be found to be spotted because of fungous growth. Nylon fabric can be turned pink by the chemical effects of the *Penicillium* fungus which can form colonies on its surface. And some microbes can metabolize the by-products formed when waste polyethylene is industrially treated to bring about a partial oxidation *before* being inoculated with the organisms. So microbes can already have some effects against plastics, and can be encouraged to play a larger role in degrading plastic waste if we are prepared to start the process off through chemical means. The careful selection of microbes which already show some slight action might enable us to raise new forms of organism which could, one day, degrade plastic waste. 2 or 3 per cent of refuse is typically plastic, a figure which may top 5 per cent by 1980, so the disposal of the unwanted surplus is going to present us with increasing problems as time goes by. Viewed in that way, it is tempting to see the extraordinarily persistent nature of plastic as a serious menace.

But this could be a colossal mistake – my view is that the non-degradable nature of plastic is no drawback at all, but the *greatest single benefit* plastics embody. Many of the features of our lives are subject to decay; wood rots, iron rusts, rubber perishes, and so forth. At the moment hundreds or thousands of miles of British gas-piping is in a porous state, because corrosion and the effect of soil microbes have attacked the iron from which they were made – it is, in effect, returning to an ore-like oxidized state again. Fatalities have occurred since natural gas was introduced, partly caused by the increase of pressure in the distribution network. Yet still steel pipes are used to transmit natural gas, and the wrapping of hessian and tar which they traditionally have is no long-term answer at all, for many such molecules are actively metabolized by microbes in the soil. Iron, then, is not so long-lasting after all. Electrical insulation, which was traditionally a use for rubber, has raised similar problems. There have been many serious accidents caused by rubber perishing over the years. Drain pipes and water-mains fail for similar reasons; and so do many structures traditionally made of wood.

These are the instances – where durability matters above all – in which we should be using plastics. They are easier to install and to work, easier to fit together, and would last virtually indefinitely. However, the popular use of plastics today for such purposes as the manufacture of plastic bags, which consumes 100,000 tonnes of high-density polyethylene each year, is misguided. In packaging we need as little extraneous material as possible, and all of it should be – *must* be – degradable. Paper, cellophane (which is a form of processed cellulose), and card are the obvious substances to utilize. Here is another example of the way in which commercial pressures have distorted real priorities. Plastic producers seek ever more expanding markets for their products, but – since plastics are the most durable materials known, and by far the most widespread product of the petrochemical industries – their uses should be more limited. The crude from which plastic is eventually manufactured is too precious to waste, and it ought to be conserved. If we are more cautious and realistic we can see that our reserves of crude oil amount to a feedstock for an industry which can produce uniquely long-lasting and reliable components – the only ones in the world so durable – and that the attempt to make them deliberately degradable is short-sighted.

If we are not to make plastics degradable, what alternative means of disposal are there? The policy I have in mind would make the amount of waste plastic considerably less than it is now, of course, and the residual burden could be disposed of through three channels. Some of the plastic waste could doubtless be burned. Many polymers are rich in stored energy, and their use as a fuel could certainly be considered. Plastics that do not burn readily, or which emit smoke, are exceptions – but the combustible forms would allow us to release some of the energy contained in the original petrochemical which would otherwise remain unexploited.

Microbe Power

A second means of disposal is recycling. Plastics which melt with heat could be used to produce crates, palettes, or other components in which the colour or the purity of the final article is immaterial. And they could go to make buried structures too, which would be uniquely able to resist the effects of corrosion in the soil, or protective undersea 'housing' for fish spawning areas. They could also make fermentation chambers or sludge units for microbial waste processing.

Finally there is the possibility that plastic waste could be shredded and mixed with compost or soil as a conditioner. This seems at first sight like the kind of idea which would cause pollution. But there are many durable constituents in the ground, fragments of stone for example, and small particles of plastic would fit into the niche already occupied by many of the natural solid components of mature soil. What we have to realize is that earth is a kind of natural spoil, containing inorganic matter of innumerable kinds of organic remains at all stages of recycling. A small percentage of another inert constituent would not matter in the least.

Rather than try to make plastics as vulnerable to decay as most other materials are, I think we should recognize them as a uniquely durable group of constructional materials with special, valuable applications. And in the event of a petroleum shortage plastics could even be made from carbohydrates obtained from growing plants; so they can continue to be with us even if oil supplies dry up altogether.

Domestic and industrial wastes have been seen as a problem for too long. With the microbe on our side, refuse could be seen as a valuable commodity – an untapped source of solar energy, ready for use by man; and effluent can be regarded as a culture medium. Microbes and the energy of life can help us overcome many of the problems which have no industrial answer. The fact that these most minute of living things can undertake tasks which the mightiest industrial complex fails to master, can teach us something about man's desire to generate power, almost as a corporate symbol of virility. And the fact that we have not looked to the microbe for assistance on any large scale in the past shows us how blind we are to its advantages and how biased we are against it.

5 Microbe Ecology

Two decades ago very few people knew what ecology was; now not only is the word familiar on a comic-strip level, but its original meaning (which concerned the interrelationships of living organisms and their environment) has broadened until now it is quite in order to talk about 'the ecology' as though it was not a study at all, but a substance. Ecology seems to concern itself with combustion and pollutants, with decreasing reserves of raw materials and hard-pressed supplies of food – but these are really more recent points on which it is fashionable to focus attention. Real ecology is not founded on mankind's technological overkill at all, but on nature; and the prime mover of the earth's ecological networks is the microbe.

Perhaps the most vivid and accessible example of microbe ecology I can quote is also the first example I came across. As a schoolboy in a garden-hut laboratory, I stood a jar of the pondweed *Potamogeton* on a window-ledge. It grew well enough as it stood in the sunshine there, but when left on a bench away from the light it began to turn, as I thought, mouldy. Small tufts of a wispy grey growth appeared around the leaves. I looked at them closely, and noticed that the tufts of 'mould' twitched spasmodically when disturbed. Under the microscope I saw that the growth was not of a mould at all, but consisted of minute bell-shaped cells suspended on delicate, spirally wound stalks that contracted at intervals like a coiled wire spring. This organism was *Vorticella*, well known to pioneer microscopists of past centuries as the Bell Animalcule. Around its open bell mouth it bears a few rows of fine beating hairs, or cilia; and they conduct a spiral stream of particles into its mouth opening. From time to time the stalks contract, dragging the suspended cells back to where they are attached, before they gradually open out again and repeat the process pointing in a slightly different direction. The *Vorticella* microbes are like vacuum cleaners, hoovering their food out of the still water.

What was happening was that the pondweed leaves, cut off from their sunlight energy and dying, were gradually decaying. Bacteria and other small microbes were removing the leaves' cells and recycling their contents. The *Vorticella* colonies had sprung up to mop up these microbes, so that what was happening was that the constituents of the pondweed were being systematically reprocessed into new living *Vorticella* cells. What impressed me most of all, apart from the indescribable beauty of *Vorticella* itself, was

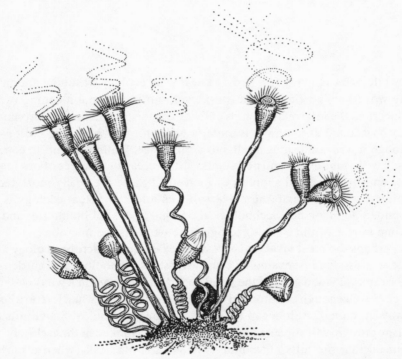

Fig 9: A colony of *Vorticella*. Three of the individuals are in the contracted state; two others are seen in the process of extending their anchoring spiral stalks; whilst one on the right is dividing in two. Though these microbes seem to be of immense importance in purifying drinking water for mankind, as recently as 1972 the living components of filter-beds and their role in purifying water were still unexplored.

that the cycle was reversible. If the jar was returned to the window so that the pondweed had a chance to obtain new solar energy supplies, and so to revive, its new lease of life meant a drop in the supply of materials to feed the microbes, and the *Vorticella* colonies seemed to become less marked. In all probability they were now being starved out of existence – and their own chemical remains might well have returned to the plant, to be incorporated in its freshly regenerating structures.

A sequel to this simple demonstration was that it seemed to explain a way in which water could be purified by microbe activity. Indeed I later was given some samples of filter-bed sand from a water works. They showed heavy growths of *Vorticella* and related organisms. So this was probably how the water was being purified: unwanted microbes were being swept up and consumed by colonies of ciliates attached to the sand-grains. Little wonder the process was so successful in making water clear for drinking. I was so impressed by the efficiency of the process that I subsequently included a detailed description of it in a pilot school's programme for the BBC in London. That was in 1962: and it is perhaps an example of how indifferent to the protozoa we have been that, ten years later, it was still being said in the journals of water research that nothing was known about the mechanisms by which these organisms in filter beds operate. The announcement that *Vorticella* played an important role was officially made as this book was going to press, which suggests that, though schoolboy experiments are no substitute for laboratory research, they can sometimes produce results that anticipate the high-powered (though sometimes inward-looking) activities of large research organizations.

Vorticella is a mainstay of civilisation. Mankind likes to live in condensed groups, whether cities or villages, and we rely on *Vorticella* to remove from our water supplies the impurities and the disease germs that we so often spread around. We take pure water for granted, as often as not, yet it is the fundamental need of all societies – and it could not be obtained without *Vorticella* cleaning it so efficiently. After an earthquake or some other major disaster the main fear is an epidemic. Why? Solely because people may then be drinking water that has not been through the microbe treatment. The damaged water-mains short-circuit the cleansing mechanism. Everyone acknowledges how disease germs spread. But how often do we reflect on the way in which microbes like *Vorticella* normally achieve something far more miraculous: the purification of water on a vast scale? *Amoeba* (which we are all taught about) is of no great ecological significance, but *Vorticella* is vital to society. When the study of its activities is incorporated into every school syllabus (and as I have explained, it is nothing complicated), perhaps we will begin to understand how much mankind owes to these truly magnificent microbes.

Now these experiments are simple enough. But they do remind us of the important fact that microbes are the power behind the ecological cycles which we see all around us, and yet the growing debate about ecology has hardly concerned itself with microbes at all. People argue about the amounts of sulphur dioxide emitted by our factories and power-plants as though this was an exclusively man-made addition to the atmosphere, yet the microbes that break down seaweed around our shores generate more sulphur dioxide in a year than all the industrial sources combined. On the few occasions that microbes do creep into the ecological argument it is as an afterthought, and sometimes as though microbes were not even alive but merely some form of basic 'chemical substance' or other. A recent ecology-minded book even goes far as to state: 'Bacteria is a potent factor . . . it gives the soil the power to digest.' '*It*', indeed!

When microbes do come into the picture the effect they have on the discussion can be profound. If we read of thousands of gallons of effluent containing cyanide or carbolic acid running into an estuary then it is easy to feel horror-struck – until we remember than many microbes metabolize these compounds and thrive on them as a staple item of diet. Small concentrations of these 'poisons' are an asset to many microbes, and not necessarily an ecological burden at all.

All the time we argue about the tonnes of pollutants emitted by industry we are ignoring the natural processes which recycle the vast tonnages of wastes produced by nature herself. For instance, it has been shown that the 6 million tonnes of anchovies which exist off the Southern California coast produce each day as much excrement as ten cities the size of Los Angeles. When we recall that the anchovy is one of thousands of species of marine life in the area, it is easy to see that the biological justification for removing all sewage from waste water that is to be discharged into the ocean is dubious. There is a strong aesthetic motivation, which it would be wrong to overlook; and if serious diseases break out then there is a public health reason to purify effluent before it is released. But the whole idea (and it is a very widespread one) that the turbid sea you can observe in some parts of the world is due to sewage from humans, is out of balance with the reality of nature.

In areas where discharge of effluent takes place from the end of a submerged pipe, one can obtain a fairly good index of the biological effect by sampling the sea-floor mud and measuring the worm populations. Since it is microbes that feed on the effluents directly, and higher organisms (such as worms) that feed on the smaller creatures, an increase in the life-support systems of an area of sea-bed is reflected in the population of worms. A typical figure would give 200-300 grammes of polychaete worms per square metre of sea-bed, which can easily double (and sometimes treble) near an outfall.

I have referred to the limits that are placed on a lake's capacity to absorb pollutants, particularly the need for dissolved oxygen in quantities that are

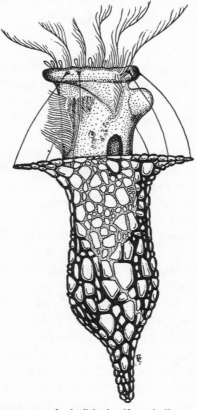

Fig 10: *Tintinnopsis campanula* builds itself a shell out of small sand grains. It floats out to feed, suspended by several anchoring threads, and jerks back inside the protective mantle if disturbed. The mat of cilia on the side of the cell (on the left in the diagram) seems to be used to sweep waste matter from inside the shell. There are almost 1000 species of *Tintinnopsis*, and they are common in the sea. None the less they lack detailed investigation.

sufficient to oxidize them through microbe activity. These limitations do not apply to the sea anything like as much. The oceans are fairly well oxygenated and, if anything, they are short of the chemical raw materials of life. There is no doubt at all that discharge of effluent and sewage far out to sea, and on the sea-bed, would not only be a useful aid to over-stressed human communities but would provide the oceans' communities of organisms with some much-needed nutriment. None of this is meant to imply that the strictures that have been placed on marine pollution are undesirable, or that we should encourage dumping on a large and unchaperoned scale. But it does put our contribution to the waste burden into perspective – and it underlines the need for us to discover how microbes work, and how great is the scale of their activities.

The Oklawaha River project in the United States shows how seriously we ought to take microbe ecology. The idea first arose in the years of the Second World War, when the possible threat from Nazi U-boats led the military planners of the United States to look for a means of moving ships and barges across the northern neck of the Florida peninsula. Plans for a canal were drawn up, which involved an extension of the Oklawaha River. Bureaucracy being what it is, funds for the projects were eventually allocated in 1964, when one imagines the threat from Nazi submarines must have considerably diminished, and the scheme began to encounter difficulties almost at once.

The project was modified to include a large recreational lake. A vast tank-like tracked earth-mover was brought in to level the trees on the lake-bed. All the vegetation was crushed into the ground, and the surface was left level and smooth, ready for flooding. So far the idea seems sound enough – but shortly after the flooding was completed, the new lake ran into trouble. The broken trees began to rise, ghost-like, to the surface and floated there, rotting away. The dead vegetation on the bottom of the new lake was degraded by microbes and large amounts of nutriment were released into the water in consequence. Overgrowths of aquatic plants (such as the water hyacinth) soon appeared, and these were quickly controlled by the liberal use of selective weedkillers such as 2,4-D. These killed the plants, but caused a build-up of decomposing plant tissues which soon rendered the water putrid and foul. As an amenity, the new lake was a total failure – the only real result of the project had been to destroy a site that had hitherto been largely unspoiled by man.

The project had failed because due attention had not been paid to the microbe. The foul water resulted only from attempts by the microbes to redress the damage wreaked by mankind. If the trees had been felled and harvested, instead of being ploughed into the ground, the soil might have stood a better change of adapting to its new, submerged environment. If the water hyacinth had been dragged out of the water, instead of being killed and left to rot, then the chemical constituents in the plants would have meant a net loss to the overloaded water instead of becoming an insurmountable burden. In future, no such project should be undertaken without a review of

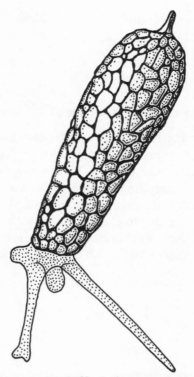

Fig 11: A shell-building amoeba, *Difflugia*, is very common in a freshwater ecosystem. How a 'shapeless' jelly-like microbe contrives to build itself a characteristic protective covering from detritus in a pond is open to debate. New projecting pseudopodia are sometimes produced very rapidly, the cytoplasm inside swirling round almost like smoke. Sometimes a single pseudopodium is produced, and the organism can then crawl along surprisingly quickly, the cytoplasm circulating like the caterpillar-tracks on an earthmoving vehicle.

the microbe's role. Desparately little is known about these vital organisms, and further research is in my view a matter of urgency.

A mainstay of biology teaching and the orthodox view of ecology is the food chain. It is a concept which underplays the role of the microbe, and by presenting the complex network of food and feeder as a series of linked boxes the food chain does not merely simplify ecological relationships but distorts them almost beyond recognition. Perhaps the most direct illustration of the way microbes have been left out of the proceedings is the typical food-chain diagram which sums up man's relationship with other organisms. The example chosen here is lucid and – as food-chain diagrams go – comprehensive. The principle omission (which has been rectified as shown) was

89

that of the land-plant link, which students of biology are taught in the following form:

either
PLANTS ⟶ (eaten by) ⟶ MAN
or
PLANTS ⟶ (eaten by CATTLE) ⟶ (eaten by) ⟶ MAN

Even this basic representation of the way man feeds creates a false impression without reference to microbes. It is microbes that provide the raw materials for the plants to grow, partly by fixing nitrogen in a form accessible to plants, and partly by recycling organic residues so that their chemical constituents are ready for the plant roots to absorb. It is microbes in the cow's rumen that make it able to convert its grass diet to protein; and it is microbes that process many of the foodstuffs before man eats them. We have touched on basic, traditional foods such as bread, wine, and cheese earlier

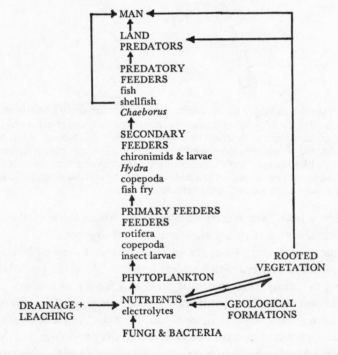

Diag 4: A typical food chain. This diagram summarises what is generally held to be the relationships between feeder and foodstuff. Note the relatively unimportant role ascribed to microbes.

on, but many less familiar dishes, such as sauerkraut, come into that category.

The more detailed 'food-chain diagram' (see p. 90) compounds the errors. To begin with (starting at the bottom) it is not only *fungi* and *bacteria* that are responsible for the provision of nutrients. Other forms of microbe (algae and protozoa) are involved. The *geological formations* are released into a form available for assimilation by many processes, and it is undeniable that the microbe plays a leading role in some of them. The capture and recycling of *drainage* compounds in *run-off water* is principally due to microscopic algae (such as the diatoms discussed on p. 123), and without those microbes much of the nutriment material washed out of the land by rainwater would be lost to the ocean depths. Moving up the diagram a little, we can see the immediate importance in the growth of *rooted vegetation* of a proper microbe community, both in the maintenance of health and in the provision of assimilable nutrients. The *phytoplankton* in the diagram *are* microbes, of course, and without them all the higher animals (from you and I to shrimps and worms) would be denied their normal sources of food.

Microbes play an important part in the provision of food for the *primary feeders*, for they capture the solar energy and trap it in chemical form and this is the fundamental source of 'fuel' for all the organisms in the chain. But there is a return loop here, too: all the excreted material from the primary feeders is broken down through the activities of other microbes and are in this way recycled through the system. Many of the *insect larvae* contain microbes without which they cannot mature properly, and all their waste matter (together with their dead bodies in due course) are disposed of in a similar fashion. Microbes inhabit the gut of the *predatory feeders* and the *land predators* too, and throughout all these types other microbes are at work preserving the health of the individual (a topic discussed in the chapter to follow).

Let us therefore put the microbe into the picture. At once the importance of micro-organisms becomes clearer so that we can see them as the fulcrum, the prime mover almost, of the whole system. They motivate the many routes through which wastes are recycled, and draw our attention to the return route through which wastes re-enter the system. What we dub waste may be such to the species which exretes it, but faeces and other forms of excrement are often heavily populated with microbes already using them as a food material. In any event, organic wastes of this sort are rapidly metabolized by microbes occurring in nature so that – though they may seem like an unwanted by-product – they are a staple item of diet for these

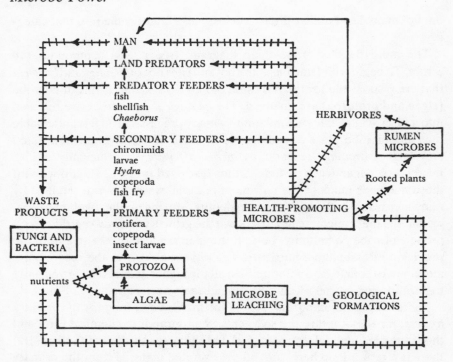

Diag 5: The modified food chain. We can see that the new pathways (represented by the hatched lines) and the proposed role of microbes (in boxes) reveal the motive power of the system and give microbes a fairer share of recognition.

smaller forms of life.

The conventional 'food chain' is easy enough to formulate:

Insects eat plants: birds devour insects: predators catch and eat the birds . . .

or:-

Worms consume vegetation: fish eat the worms: mammals feed on the fish: predators finish off the small mammals . . .

and so on. Everyone is taught them, but the 'food chain' distorts what happens in natural ecological interactions. The analogy is too simplistic and admits none of the subtleties and variety of life as it is lived – almost as though one drew an analogy between a football game and a pin-ball machine. The 'chain' shows how one organism in the system devours the next in line, and is itself eventually eaten by the following creature in the list.

But the food system functions in a far more subtle way: for instance, it is relatively unusual for a given species to feed only on one prey. So typical 'chains' divert our attention from the fact that the interrelationships are more like a branching, three-dimensional network. The expression *food web* is nearer the truth.

The 'chain' also prevents us from appreciating that the quantities vary enormously from one level to the next. It is not, for example, that a duck eats a worm and is in turn eaten by an eagle. The duck will eat thousands of worms; and the eagle will need scores of ducks to survive. Instead of:

worm ⟶ duck ⟶ eagle

we should have:-

5,000 worms ⟶ 35 ducks ⟶ 1 eagle

The concept that a large animal eats a great many smaller ones is now termed a 'pyramid of numbers', which goes some way to acknowledging the purely quantitative concentration we observe.

Yet even this is an absurd oversimplification: eagles do not only eat ducks, and ducks eat other organisms besides worms! So the system is a branching, dichotomous pattern of pathways more like a three-dimensional family tree than a 'chain', with branches that coalesce like tributaries in a river. You can show, for example, that a young herring feeds on two dozen different types of organism, and any one of those feeds in turn on eight or ten smaller creatures. This important quantitative disparity, both of species and numbers, accounts for the way in which cumulative toxic materials can accumulate from one level to the next; and explains why I far prefer to speak of *food funnels,* rather than chains. In the 'chain' idea, DDT can be represented as passing from one link to the next, and there is no indication of the insidious accumulation which one can observe in nature. In the alternative food-funnel concept we can demonstrate with graphic clarity how the open end of a tapering funnel can embrace a great number of food sources and concentrate them down to the apex, where the accumulation is most pronounced. By overlapping the funnels we can further reveal the interdependence of feeders on their food, and the role of dietary substitution in the event of one constituent being in short supply.

We can revise the traditional nitrogen cycle to involve microbes, too. This shows how atmospheric nitrogen and nitrates produced by decomposition in the soil, are synthesized into protein by the green plant. That protein is

93

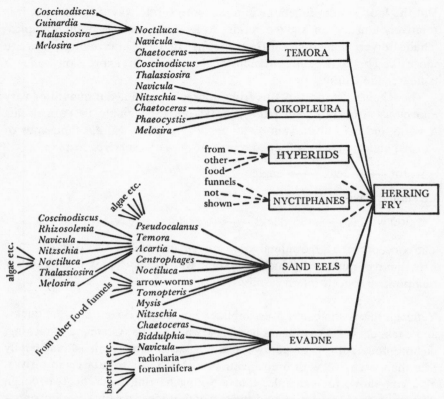

Diag 6: A proposed 'food funnel' system. This concept seems to be a sensible way to understand the relationships between the different levels of predation. Even this complex diagram reveals only a small part of the complexity found in nature: a three-dimensional model would be necessary to illustrate the whole concept.

excreted by the animals which eventually consume it, and they themselves may undertake (or undergo) predation; or alternatively it returns to the soil when the plant dies. In any event it ends up as waste matter, ready to go round the cycle again. The importance of microbes in making nitrogen available, and in converting one form of nitrogen-rich matter into another, is poorly emphasized in the orthodox scheme. At all the levels of recycling, microbes are available to reintroduce fixed nitrogen to the system. And of course the way animals feed on microbes is omitted altogether. Not only does this explain the vital way in which cattle use microbial colonies to synthesize available protein out of digested vegetation, but the direct route to man himself is likely to become important as cell food enters our diets on a larger and larger scale.

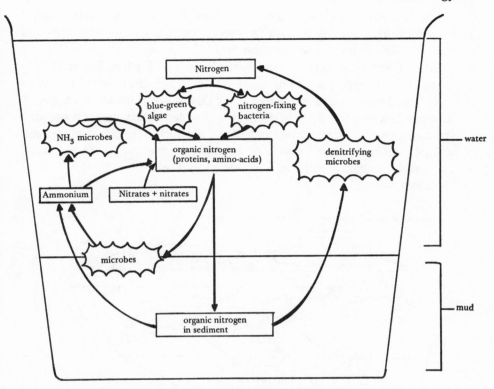

Diag 7: The movement of nitrogen in a freshwater lake. This condensed version of a small part of the 'nirogen cycle' shows how fundamental are the processes involving microbes (in starred boxes).

The microbe-modified nitrogen cycle has several extra loops that do not generally feature on the orthodox diagram, and a few that are never included. But it is the 'carbon cycle' that is most in need of rethinking. As it stands in the textbooks of today, it is based on the movement of carbon through carbon dioxide (CO_2) in the air until it is captured by green plants and stored in the form of carbohydrate. These die (or are eaten by animals) but in any event they are eventually broken down by bacteria and fungi so that the carbon dioxide is returned to the air. The gas is also released by the various living components of the cycle as a result of their respiration. It is a system reflected in what one is taught at school: 'you breathe oxygen IN, and you breathe carbon dioxide OUT'. This is a distortion of reality: in truth we breathe out a little more CO_2 than we have breathed in; and we breathe out a little less oxygen than we have inhaled. Basically, this is true of all respiring

95

organisms, including green plants. The oxidation of foodstuffs is the energy source of life, and respiration in green plants relies on the principle too – burning fuel through metabolism.

The 'carbon cycle' does not involve only carbon. Each molecule of CO_2 taken in by a green plant to be made into a carbohydrate is united with a molecule of water, and each molecule of CO_2 lost in respiration corresponds to water loss too, so we could draw up a complementary diagram showing the movement of water in and out of the chemical systems of plants and animals that is very like a mirror of the 'carbon cycle'. So the system is not exclusively concerned with carbon: and indeed, is it a 'cycle' at all? No, I think not.

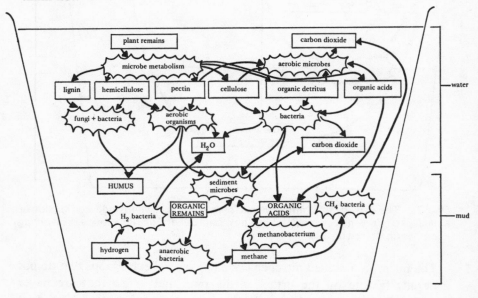

Diag 8: The movement of carbon in a freshwater lake. When plants die, their remains are subjected to a complicated series of processes before the carbon dioxide and water from which they arose are returned to the environmet. Microbes carry out these conversions, which could provide a model for industry in the future. *(cf Diag 9)*

As the life processes of living things rely in some way on oxidation, there is a need in the system for the initial, primary reduction to take place. That is to say, there is clearly one site at which the energy is put in to the scheme. This is photosynthesis, in which green plants absorb light and utilize the energy to combine the water and carbon dioxide in the first place. The weight of all living things on earth (the so-called *biomass*) results entirely from this primary reduction using light energy, with small exceptions such as the microbes that exist by oxidizing inorganic chemicals (and many of those compounds are produced by other microbes anyway). The splitting of water by photons – light particles – and the reduction of carbon dioxide to carbon, results in a long series of reactions as one organisms eats the material substance of another. The sun shines on a meadow and is trapped by grass leaves; the grass is eaten by a cow; the intake is digested by microbes; the microbes are assimilated by the cow's digestion; the cow is eaten by a wolf; the wolf by vultures; a vulture by rats; the rats by an owl; the owl by an eagle . . . and so on. At each stage some of the energy is released, and the carbon dioxide and water that result from the oxidation are returned to the environment for recycling. And eventually we find that the dead organic remains in the soil are eaten by worms, by protozoa (discussed in the following chapters), by bacteria, and in the last resort by fungi, which can extract the last dregs of stored energy from the system. On average the recycling time is perhaps two centuries, and there are at any one time many millions of billions of tonnes of biomass on the earth's surface. (The figures are 3×10^{12} megatonnes of plants, and 1.5×10^{12} megatonnes of animals; some 10^{18} kilocalories of sunlight are involved each year in the primary reduction reaction).

Throughout their lives, then, living organisms are transforming hidden chemical energy into the energy of metabolism, but with one important exception: the moment at which light is trapped and converted into chemical energy. This is no 'cycle'. It is an initial stoking-up; it is a priming of a system with energy, which then runs down until the energy is gone. At the end of the long train of events we are left with the simple molecules of carbon dioxide and water and the only way the system can be restarted is through the energy of the sun acting on a green plant. It is like the train of a Big Dipper, hoisted to the top of its run. The powerful store of poised potential energy is capable of being wasted (and in this analogy the burning of a log is like allowing the Big Dipper car to run straight back down to its starting point – which would be a short ride indeed, and no fun for anyone), but if we are wise we can design the run so that the energy comes out in slow, controlled stages.

This Big Dipper theory is a departure from the conventional view, but I have proposed it for a specific reason. Raw materials derived from green plants are replete with energy, and in a society which is short of energy (and which pollutes itself when it has adequate power supplies) it is high time we arranged that this inexhaustible supply of power was harnessed and guided into useful channels rather than being wasted wholesale. No commercial Big

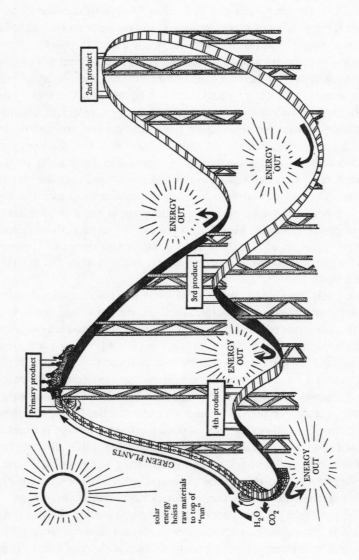

Diag 9: Energy conservation analogised by the 'big dipper' model. Solar power is converted into an energy store by green plants. The production of successive useful materials by cell technology enables much of the energy to be harnessed.

Dipper would work if its train simply ran back at full speed down the steep slope up which it had just been laboriously hauled. But that is what happens when we burn fuels. If we were a little more thoughtful and long-sighted about civilization we could ensure that we built into the Big Dipper's design

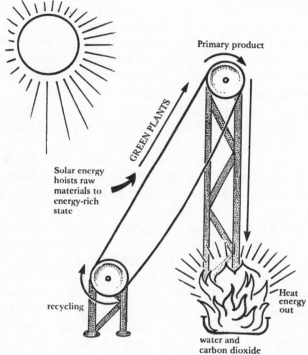

Diag 10: Energy wasted by combustion. The painstakingly elaborated solar energy store of the 'big dipper' is wasted in combustion as happens when we burn wastes and liberate heat. Yet even this primitive attempt to harness chemical energy in refuse and waste is regarded as relatively advanced.

the maximum number of successive peaks, each a little lower than the last, but representing the efficient use of the sun's unceasing bombardment of our plant with free and illimitable energy.

If there was a 'cycle' our planning would be simpler. The chain of events would be self-perpetuating, driven endlessly round by the sun. But the only use of the sun is to charge up green plants with stored energy – to hoist that Big Dipper back up to the highest point on the slope. After that, a veritable spider's-web of alternative pathways exist; in the past it has been due to chance, and to nature, what happens then. In future it should also be up to man.

Microbe Power

Plotting what lies ahead for our earth environment has become increasingly popular in recent years, and every climatological trend has been monitored with interest. Unfortunately the enthusiasm has not always been matched by constructiveness, and as a result we have been left with a series of conflicting prognostications. The classic example must be the temperature paradox: on the one hand, it is argued that carbon dioxide levels are rising owing to the activities of industrial firms, so that more heat energy will be trapped and the climate will become warmer; on the other hand, we learn that the same fuel-burning industries are adding to atmospheric dust, and that will restrict heat intake so that the temperature will fall.

Empiricists will say that this is perfect, since it leaves us exactly where we want to be. Cynics will say that this exemplifies the lack of scientific understanding of the subject. But the realist might divert attention to what changes might be taking place anyway – and it is certainly true that our obsession with wanting to see things stay as they are is very far from what nature likes to provide. The amount of carbon dioxide and dust we produce through our industrial activities is trivial compared to the activities of living organisms, and we are losing sight of the fact that the earth, bathed as it is by incessant nuclear radiation from the sun, is subject to enormous influences which we do not yet understand. Our planet has always tended to waver from one extreme to another. For most of its past history, for instance, it has been more tropical than it is at present; for ninety per cent of the time there have been no ice-caps at the poles. The Ice-Age cycle of more recent geological time shows how much the earth is still changing, and will very likely go on doing so. Some recent data have suggested that the mean temperature over the globe has fallen very slightly over the past few decades, which is just what the temperature has done for long periods in the earth's past, and yet our insistence on looking at ecology from the wrong angle makes us search out an industrial reason for this – even if natural fluctuations are greater.

Do not misunderstand what I am saying: this is no argument in favour of rampant energy-consumptive expansion, but it does suggest that if our climate does wander we must look at the new trend in the light of the huge changes of the past. We have had those ages of vast shallow seas, where microbes built up the immense beds of chalk, limestone, and diatomite we see today. We have seen ages of gigantic plants. Did they perhaps flourish because levels of carbon dioxide were higher then, as a result of the previous aeons of volcanic activity? This would have meant that temperatures were higher, too, and this in turn would have triggered off widespread plant growth which would have greatly increased the amounts of oxygen in the atmosphere. I suppose that could have been a factor in the development of the giant reptiles which followed – but, in any event, the churning cycles of tropical forests, glaciers, lakes, and the drifting continents mean that we can hardly hope for the earth's climate suddenly to stand still in time for the sake of our peace of mind.

I have even read one report which concluded that the removal of growing trees, and the consumption of atmospheric oxygen by fuel-burning, would necessitate the manufacture of oxygen in future by machines designed to keep our air breathable! But both aspects of this alarming allegation are erroneous. As the recently published *Study of Critical Environmental Problems* pointed out, if all the fossil fuel in the earth's crust was burned, it would result in a lowering of the oxygen content of the air from 20.9 per cent to 20.8 per cent – a trifling amount of change. Even if that impossible end was attained, the effect would be counterbalanced by microbe activity. More carbon dioxide in the atmosphere (produced by the combustion) would increase the growth rates of plant life with a corresponding rise in the rates of oxygen release. The current arguments often state how important trees are in providing us with our oxygen supply, but here too the microbe's activities are ignored. Surprisingly, perhaps, as much as 90 per cent of our oxygen comes from microbes. If all the world's trees stopped producing oxygen tomorrow it would be difficult to measure the effect, but if the microbes gave up that task we would soon know the difference.

Oxygen is not the only gas in the atmosphere produced by microbes. They are even responsible for some much-publicized oxides of nitrogen in the upper atmosphere which are said to be a threat to the protective ozone layer. The NOx diffuses through the atmosphere, and a measure of it eventually ends up in the stratosphere. Opponents of supersonic jet aircraft have been quick to point out that NOx occurs in the exhaust gases of *Concorde* and could damage the ozone layer which occurs in the uppermost region of the earth's atmosphere.

Perhaps 'layer' is the wrong word for it. In thickness and concentration, the amounts of ozone vary greatly from place to place and from time to time, so that daily fluctuations of 25 per cent are quite normal. And the idea that the NOx would damage the ozone seems to me quite wrong. The ozone has formed through the intensely energetic solar radiation literally breaking up and rearranging atoms of oxygen from the normal molecular form (O_2) in which two atoms make up one molecule to the ozone configuration (O_3) – a triad. Even a non-mathematician's reaction to the notion might be that the amounts of energy involved are so colossal as to make the ozone 'mantle' seem self-perpetuating; but the critics reply by arguing that laboratory models show how NOx can attack ozone, and therefore we must avoid releasing even the smallest amounts of NOx into the atmosphere.

Now let us introduce the microbe into the argument. NOx molecules are not only produced by man-made reactions. I am not thinking of oxides of nitrogen emitted by volcanoes and fumaroles, either, but the relatively large amounts produced by the microbes of decay. The threat of NOx is hardly novel; bacteria have released it for thousands of millions of years, and they contribute a very large proportion of the NOx now being detected in the atmosphere – unfortunately we cannot distinguish which molecules are 'natural' and which 'synthetic', but the mere presence of NOx.is nothing new.

It is interesting to note that *Concorde* has already had to heed some advice from microbiologists, who have pointed out that wet or contaminated aviation fuel could act as a substrate for microbes. While the cells were growing in fuel tanks (doing nothing more sinister than mop up after mankind's imperfect cleaning of the fuel), they posed a potential threat to the safety of the aircraft, since they could clog fuel ways, And now, as though to redress the balance, microbes in the soil can show how NOx can coexist in nature with a stable ozone mantle. This not only shows that NOx is less novel than we might have imagined at first, but it also suggests that oxides of nitrogen released by *Concorde* might be less of a threat than originally feared.

The importance of the ozone layer is its ability to filter out most of the harmful ultraviolet wavelengths that bombard the earth from the sun. Without the filtering effect, we would suffer more radiation sickness than we do (in the form of sun-burn, which is actually a radiation burn, and also skin cancer). This is a side effect of circling a few million miles away from an active, self-sustaining nuclear reactor; and we are fortunate indeed that the levels of sunlight that reach us are so delicately balanced as to weigh risk against energy intake.

An even greater threat to the ozone has recently been put forward, namely the chlorofluoromethanes (best known as freon) from aerosol sprays and refrigeration units. The idea behind this latest proposition is that organic compounds of chlorine can be stable enough to reach the stratosphere, and cause a catalytic breakdown of the ozone (the chlorine atoms being regenerated, and so becoming available to initiate a vast chain reaction). But here too the argument has omitted to consider organic chlorine compounds from microbes. One of them, methyl chloride, is present in the air at far higher levels than freon and has been with us for millions of years as a result of microbial activity. No adverse side effects seem to have accrued. And one can apply a similar analysis to other pollution problems.

Sulphur dioxide is an atmospheric pollutant known to exacerbate bronchitis, and attempts to reduce levels in industrial areas have been successful in many parts of the world. But sulphur dioxide is only dangerous when it is present in abnormally raised concentrations, for this gas plays a vital role in the natural scheme of things. The amount of the gas produced by industry is much smaller than the quantity produced by microbes. Sulphates are a vital nutrient for plant life, and the continuous leaching out of sulphate from the high grasslands by rainfall would make life on high moor and scrub very difficult indeed. These upland plants rely on sulphates released by the sulphur-containing rainfall for much of their supply. Even when we consider this gas – this well-recognized pollutant – we ought to make allowances for the amounts produced by microbes, and also for its important function in *maintaining* life.

A similar argument could be applied to carbon monoxide. The interpretation of these levels should, I am sure, take into account the monoxide which microbes produce. The microbe explains some of the accounts in recent years, which have shown how levels of carbon monoxide (which has always been assumed to be principally the result of man-made combustion, particularly in motor-cars) seem inexplicably to fall. Where is the gas going? Undoubtedly a leading 'mopping-up' agency is the microbe population, many members of which may play a part in oxidizing the gas. It is a mistake to leave microbes out of our discussions on 'the ecological balance' and the effect of small concentrations of man-made molecules in our environment. We will find in most cases that a so-called pollutant is already widespread in nature from microbial sources – and, more important, we can recognize that it may play an important part in maintaining the life systems of the earth. It is tempting to argue thus:

Substance X is produced by man-made processes;
In excess, substance X is poisonous to mankind;
Substance X can be detected in the environment;
THEREFORE it has to be banned.

But we can look further and may discover that:

Substance X is produced by microbes in nature;
It is vital for life processes of other living organisms;

Microbe Power

Substance X has always been present in the environment;
THEREFORE, though excesses which might be a hazard to health must certainly be ended, the substance itself is an important thread in the web of ecological interrelationships which has been built up over millions of years.

These are the problems which man, the newcomer, has yet to understand.

6 Microbes for Health

It is difficult to start a discussion on the microbes as the bringer of good health when our instincts are so heavily weighted against the idea. The everyday concept is that germs are microbes, and that germs make us ill; so we keep well away. But the association between *germ* and *illness* is relatively recent. The term more accurately signifies a kind of living essence, and it derives from the Latin *seminaria*: this is why the *germ cells* are found in the *seminal fluid*. Some of our expressions remind us of the time when the concept of a germ had more positive connotations, for we still speak of 'germination' and of the 'germ' of an idea.

The idea that germs were essentially bad become popular in the late nineteenth century, when Louis Pasteur energetically championed the germ theory of disease. It is just as well he did, for the result was a steady improvement in the standards of cleanliness and hygiene that helped to rid us of many traditionally feared infections. Unfortunately, the legacy of the theory has been to turn public opinion against microbes in general, and as we have seen this has led us away from many important answers to urgent problems.

The dawning of the concept took centuries. Bacteriologists who are asked for an example of a scientist who awoke earlier to the nature of bacterial infections will probably cite Robert Koch, the country doctor whose hobby it was to culture microorganisms and who described the organism of anthrax. But we can go back before him. A largely forgotten French biologist named Davaine worked with anthrax a generation before Koch: he isolated the bacteria, demonstrated them in the blood, proved that they could not pass the placental barrier to infect the foetus of a cow with the disease, and even went so far as to filter the bacteria through unglazed porcelain (proving that the clear filtrate, which was free of the organisms, was also free of the ability to induce the infection). A decade before Koch had shown how to grow bacteria on a bed of nutrient jelly, a German bacteriologist, Klebs, had cultured them on a similar medium made from fish glue. It was also Klebs who carried out the historically important induction of syphilis in a monkey, seven years before the Russian Metchnikoff who is usually associated with it.

We can go back to 1844 to find Bassi, the Italian whose name is commemorated in the pest-control microbe discussed on p. 31, writing of his belief that many diseases – including smallpox, the plague, and syphilis – were caused by 'minute living parasites, whether animal or vegetable'. In 1840 Henle wrote about infectious disease and showed that he felt fermentation to be essentially a similar phenomenon (that is, one caused by microbes), and three years before that Schwann published a series of notable observations on the role of the yeast cell in alcoholic fermentations.

A century before that time lived an Italian churchman, Lazzaro Spallanzani, who first disproved the spontaneous generation theory with a brilliant series of experiments involving heat-sterilized broth and bacteria. In the earlier years of the eighteenth century we find Joblot proving that heat-sterilized infusions of hay remained free of microbe growth, and it was in 1720 that Marten stated his belief that tuberculosis was caused by 'wonderfully minute' living organisms. In the middle of the 1600s we have the pioneer microscopist Leeuwenhoek, a Dutch draper, painstakingly grinding tiny bead lenses and watching microbes, and shortly before his time Kircher wrote rambling comments on the 'innumerable brood of worms that are invisible to the naked eye' in decomposing organic matter – comments believed by some to suggest that he saw bacteria. In 1653 William Harvey had written of a theory that contagion could 'lurk' in infected materials, producing its like 'in another body'; and a century before that the death was recorded of an Italian philosopher, Fracostoro, who propounded a view explaining disease that had much in common with the germ theory of the Pasteur era.

So there were three centuries of precedent for the idea, which gradually grew from these small developments into the formidable orthodoxy of today – it was no new, dramatic revelation. Yet not only has it left us with an erroneous concept of what microbes are; it has even distorted the popular idea of disease.

It is easy to forget that the most serious diseases which confront us have nothing to do with microbes: the conditions I have in mind are many, but include cancer, heart attack, strokes, disseminated sclerosis, schizophrenia, and obesity. If we restrict ourselves to infectious diseases in this definition, then even here we can see that the majority of incurable and widespread infections are not caused by bacteria, but by viruses (see pp. 157-160). From influenza to measles, and from chicken pox to mumps, the choice of example is wide. Two of the greatest epidemics (they have each exterminated entire civilizations in the past) are smallpox and malaria, yet neither is caused by bacteria. The former is a virus infection, and the latter is produced by a protozoan. So the widespread view that diseases are caused by bacteria is something of a scurrilous slur on some of our most important allies.

But, since this is the view that we are all taught, there can be little wonder it has such widespread acceptance. Less easy to accept is the confusion that still exists in science. If you look through a random selection of specialist textbooks, it is easy to see how marked the bias is. Three typical examples I have picked from the shelves show, in total, three sections on the protozoa, five on algae, fourteen on fungi, but the total of eighty-one devoted to bacteria. I will return to the question of the different types of microbes (and what part they play in the microbe world in general) later; for the moment the important issue is that we are easily lulled into a belief that bacteria are the only microbes worthy of investigation, and this is what we find in the typical textbook the world over. In universities, microbiology and bacteriology are regarded as generally synonymous: and bacteriologists and microbiologists carry out the same activities. This has been the case for many years (and doubtless explains why as a student I was once recommended not to bother with a career in microbiology, since all the bacteria were now known about). Perhaps it is time that the world of research moved on from the era of 'microbes means bacteria' as well.

If we begin to consider microbes as our helpers, and as the promoters of health, then the first example to come to mind is surely the production of antibiotics. The concept of antibiosis, i.e. one organism producing a substance that acts against another, dates from the 1890s, and the first antibiotic substance to be investigated was penicillin. This is a chemical produced by a mildew fungus, *Penicillium*, and the discovery of penicillin was followed by several years of complete lack of interest in the possible uses of these compounds. The evidence in favour of penicillin at the time of its announcement was more than enough to spark off interest in the mind of any adventurous research worker, but, apart from a few visionaries who did try a little campaigning, there was no real progress until Howard Florey and his colleagues began a systematic survey of all possible antibiotics around 1940. I do not doubt that the delay was due, in part at least, to our blind spot for microbes and a subconscious belief that they could not provide anything so useful for man. In the years that followed much progress was made, including the discovery that the addition to the culture medium of a waste product – corn steep liquor – enormously increased the rate at which penicillin was produced. Then streptomycin appeared, which is still claimed from time to time in America as the first antibiotic of them all, and subsequently more systematic investigations of microbes from soil and elsewhere have greatly extended the list.

The use of antibiotics has enabled us to control many of the great infections of the past, and the classic killers, which include puerperal fever, septicaemia, and bacterial pneumonia, are all virtually beaten. But if a microbe in culture can produce antibiotics, what about the microbes in nature – microbes we carry with us inside and out?

We are taught that, as long as we remain well washed, disinfected, and free of microbes, we will stay healthy – 'cleanliness is next to godliness', and so forth. Infections, it is said, occur when micro-organisms gain access to our bodies. But this cannot be the case. Organisms that cause disease are regularly present on and in our bodies without causing ill-health. Vast populations of these potential pathogens are found in our intestines and are released onto the skin surface whenever we defecate. Yet these organisms vanish and are replaced by the normal skin microbes within an hour or so, through mechanisms as yet only partly understood.

Or consider animals which lack our sense of hygiene. A dog sniffs the excrement of other dogs. It digs for decaying bones in the ground with its forepaws and nose, chews bad meat unconcernedly and nuzzles in odorous corners whenever it can. It cleans itself by licking and swallowing the microbes it has picked up. Yet it is fastidious mankind, well soaped and medicated, that has the dandruff, boils, eczema, pimples, acne, and the rest. Is it not possible that, in waging war on the microbes on our skin, we are wiping out armies of organisms that would protect us from infections, and not cause them at all? I believe that microbes on our bodies serve to protect us from disease.

To begin with, let us look at the natural population of the skin, 2 square metres in area, which covers the adult human body. Some accounts in the past have suggested that the diversity of the environment which the skin offers to its microscopical inhabitants is as varied as the earth's surface appears to larger living things. But this is an exaggeration, and perhaps more that poetic licence could admit. There is no deep water on the skin, no fresh water either, and very little running water. The surface is all keratin, in contrast to the enormous variety of materials in the earth's crust, and there are none of the extremes which are typical of a landscape.

Skin microbes exist in a moist, slightly oily environment composed mainly of protein; their natural water supply is almost entirely derived from the saline perspiration. This essential sameness means that the organisms themselves are pretty constant, too, and they are limited in type. The bulk of the population is composed of bacteria of the *Staphylococcus* type, familiar enough as the organism which causes boils and which often used to cause septicaemia. The disease-causing staphs secrete an enzyme known as coagulase, and most of the types that live on the skin are coagulase-negative; that is to say, they would not be thought of as potential pathogens. There are many other related bacteria present too, grouped together in the genus *Micrococcus* (which means, literally, 'little spheres' and is a perfect description of their appearance under the microscope), along with *Sarcina* which has the attractive habit of growing in neat little cubes of eight. In the moister regions of the armpit and between the toes grow significant numbers of rod-shaped bacteria (including *Mima, Alcaligines,* and *Herella*). And there are smaller populations of other microbes too, including occasional yeasts.

The skin surface of the scalp is more diversely populated, since the presence of hair prevents skin water from evaporating as readily and so the humidity is higher. The need for water is paramount for all forms of life, and so the somewhat more diverse population of the human scalp is predictable. The same kinds of yeast that inhabit the hairless skin abound, and one can always culture fungi of the genus *Aspergillus*. However it is hard to be sure whether they actually inhabit the scalp, or whether they are inevitably found there since the spores of these fungi abound in the air. Incidentally, there has been comparatively slight investigation of the microbes that inhabit the body's hairier regions. There is a growing, if small amount known about the types of skin microbes in general, but the degree of insight we have into the fauna and flora of the scalp is very much less even than this.

One question it is difficult to answer is: how many microbes are there? Different microbiologists have found different totals, depending in part on the techniques they used to obtain a harvest. Strips of adhesive tape which are used to remove the top layers of skin and the adherent microbes give an artificially low figure, and so does the alternative method of pressing growth medium on to the skin before incubating the samples. Other figures have been derived by taking portions of skin from cadavers and incubating them, but this could give an artificially high reading if the microbes have increased significantly *post mortem*. Taking into account the results from all the different methods, one gains the impression (and it is really only that) of

109

there being a few thousand microbes per square centimetre of open skin, rising to tens or even hundreds of thousands per square centimetre in the scalp and axilla, with a maximum density in these protected areas of around one million microbes per square centimetre.

I have referred to the harmless, coagulase-negative staphylococci which inhabit our skin, and so far the impression one has is of harmless microbes living out their lives on our bodies. But this is only part of the story, There are, for instance, a great many coagulase-positive staphs there as well, and they are known to be the potentially pathogenic types. The most frequent species to cause actual illnesses ranging from boils to septicaemia, *Staphylococcus aureus,* is commonly present. So is the *E coli* which occurs in the intestine, and *Streptococcus* of the kind which can cause sore throats. Pathogens are frequently found on the skin – but they rarely cause any actual disease. But is it possible that it is the resident microbe population which helps control the pathogens? Some microbes have been shown to exert a protective action against unwanted types, and if we search the examples out and assemble them they begin to look like formidable catalogue of health-promoting microbes.

One of the skin yeasts, *Pityrosporum ovale*, is known to prevent the growth of fungi, and it is thought possible that it is the presence of this microbe which protects against the ringworm fungus. In the test-tube at least, *P ovale* has the effect of inhibiting the growth of the causative organism of ringworm, though it is yet to be proved that the same thing happens on the skin. Many of the skin staphs produce substances that interfere with the growth of *Staphylococcus aureus*. Elsewhere it has been demonstrated that some of the skin micrococci can inhibit many disease organisms (so far the list includes *Bacillus, Staphylococcus, Streptococcus, Pneumococcus, Neisseria, Clostridium* and *Corynebacterium diphtheriae*). The production of antibacterial compounds by skin microbes has been known for a long time; indeed Florey published something on these lines in 1949 as an extension of the work on antibiotics which led to the exploitation of penicillin, and in the latter part of the nineteenth century, staphylococci were observed to inhibit the growth of other microbes.

But the conclusion that it is antimicrobial substances – analagous to antibiotics – that produce the controlling effect is not enough to explain all the examples of control that lie dotted about in the literature. In the early 1950s, a protective strain of *Staphylococcus aureus* was isolated from a nurse and some child patients in a hospital ward which had remained remarkably free from everyday staphylococcal disease. Investigation showed that the protective staph could readily grow in the nasal passages and on the skin, and when it was present the chances of contracting a staph infection were very much reduced. The protective strain was closely studied, but no antimicrobial substance seems to be produced by it. The way in which it exerts its protective activity remains a complete mystery – all that is certain is that this strain can largely prevent a staph infection from starting, and some experiments were even carried out in which new-born babies were deliberately inoculated with the protective organism.

It has similarly been known for scores of years that a patient suffering from a mild throat infection caused by certain staphs is less likely, as a result, to suffer from an infection caused by the more dangerous *Corynebacterium diphtheriae*. Perhaps most intriguing is the observation that washing seems to increase the numbers of microbes on the skin (possibly the water makes it easier to recover microbes, though one could equally argue that an actual increase takes place to protect the 'assaulted' skin), and the discovery that regular treatment of the skin with a disinfectant such as hexachlorophene increases the risk or cross-infection, rather than diminishing it.

Usually we speak of skin microbes as *commensals*, i.e. organisms that live on or in a larger form of life, not as parasites, but as harmless passengers, (literally 'eating at the same table'). I am convinced this is an incorrect designation: that the microbes are not merely innocuous, but that they positively promote the health of the individual. They are in this sense the converse of a *pathogen* (disease-causing), and I have proposed that we call them *salugens,* which means 'health-producing' (from the Latin *salus* = health).

The first time I saw evidence of the action of skin microbes against disease organisms was when I was on the junior staff of a medical research laboratory, and grew a selection of the microbes from my own skin and from that of several volunteers. The original purpose was to discover what microbes habitually turned up, but culture plates containing mixtures of species often revealed how one colony of organisms seemed to repress other colonies alongside it. More interesting is the discovery that some types *stimulate* the development of others, so it is even possible that salugens can act indirectly by assisting the growth of different protectors. Some people in the 1950s dubbed the effect 'satellitism', apparently by analogy with the way satellite communities spring up around towns. But since interference with microbial growth is usually known as *antagonism,* I prefer to think of this potentiating behaviour as *protagonism.* In any event it is now becoming clear that the skin microbes exert profound effects on each other of both positive and negative kinds, and surely we could benefit by harnessing these effects. Medical practitioners know that treatment with antibiotics can lead to a sudden overgrowth of other species of bacteria, or an infection with fungi. Could this not be because the salugens which ordinarily kept those harmful organisms under control have been inadvertently eliminated by the therapy?

Ever since the germ theory of disease first became popular we have tended to assume that microbes are only responsible for illnesses, and that we have to keep ourselves as free as possible of organisms. But even the few facts above show that in everyday life we carry disease-causing organisms with us, while remaining perfectly healthy. They even allow us to infer that a well-cleaned skin surface makes infection more likely, and not less. This is a startling conclusion. It tends to reverse the normal point of view and it suggests that there are many protective mechanisms about which little or nothing is known. And it certainly encourages one to think that microbes are actually good for us.

Salugens occur elsewhere in the animal kingdom. As we have seen, herbivorous animals, from rabbits to cows, could not survive on their normal diet without the help of microbes. Some insects have microbes living not just inside their bodies, but in the interior of certain body cells. We can show that they are necessary, and not merely harmless, by removing them experimentally. The results are sometimes dramatic. A normal larva of the beetle *Sitodrepa panicea* is 50 millimetres long, for example, but only attains this size if it has its salugen microbes within its own body cells. If they are eliminated, the larva grows to only one-quarter the normal size. Many other insects (including the familiar cockroach *Periplaneta orientalis*) have been found to develop to only half the normal adult weight if they are reared without their internal 'army' of microbes.

In the plant world they are important as well. Some species lack root-hairs altogether and can only exist in combination with a root fungus which lives partly inside the host cells, and partly outside the host, in among the soil particles. This association (which is known as a mycorrhiza) allows the plant to obtain a supply of nourishment from ground that it would otherwise find inhospitable in the extreme. How important microbes are to root systems in general is unknown, but the microbe populations around growing plants are large and many of them seem to act as salugens towards their bigger neighbours. In the 1930s some experiments with wheat showed how soil microbes could protect against infection. Wheat plants grown in sterilized soil were shown to be more liable to infection with a parasite, *Helminthosporium sativum*, than plants grown in soil rich in its normal microbe population. The tests showed that the parasite could survive perfectly well in sterile soil, but if a trace of unsterile, normal soil was added to the culture then the survival rate of the parasite dropped dramatically. One gramme of unsterile soil per pot of culture was sufficient to prevent the parasite from being recovered at all. Subsequent experiments have shown that even heavy administrations of the fungus *Ophiobolus graminis* are insufficient to reliably establish an infection in pot-grown wheat plants growing in normal soil. If they were grown in sterilized soil instead, however, the infection soon proved overwhelming.

Let us link these strands of evidence into the thread of an argument. There is something dramatic here: an inescapable conclusion that is in some ways as profound as the germ theory was in its time. We are faced by the fact that you may sterilize your earthenware pots, sterilize the soil, sterilize the air if you wish; but that may be just the way to encourage disease in your plants, rather than discourage it. You may bathe your baby in medicated soaps, anoint her with expensively advertized antibacterial lotions, even wear a face mask like a surgeon if the fancy takes you; but she needs her covering of microbes if she is to remain healthy. Our emergence from the birth canal when we are just nine months old coats us with vaginal microbes and initiates us into the existence we are destined to share with our microscopic protectors. All our fastidious attention to personal freshness, our obsessional routine of toothbrush, gargle, and mouthwash; those disinfectants, lotions, unguents, and the rest; all are based on what I am now sure is an out-dated premise.

The curse of smelly armpits is likely to prove to have its origin in the psychology of the individual who suffers from 'body odour'. Overperspiration with psychogenic overtones is the cause; and though washing may clear up the problem for a little while, as soon as the perspiration returns, so does the odour. It may well be that the skin's secretions influence the microbes that live there; and, if one's mood effects one's secretions, then in that way we may all exert some form of mental control over the microbes we carry along. Toothbrushing may be a fetish, too; there is much evidence that it damages the neck of the tooth as the years go by, and there is no conclusive evidence to show that religious dental hygiene has any prophylactic effect against decay. The complex bacterial associations in our mouths are part of our own internal ecology, and I do not doubt that we need them. It is quite untrue that you need to repeatedly wash your hair, or your face, either, to keep it healthy. And it is equally fallacious to suggest that all-over antimicrobial warfare is necessary for the preservation of a sweet and pure skin.

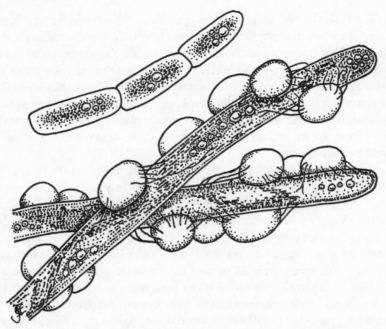

Fig 12: Microbes from our teeth. The mouth supports large populations of bacteria which must serve to protect against disease. Of particular interest are the thread-like forms which are typically covered with smaller, rounded species. Exceedingly fine threads seem to be present, holding the two types closely together. The role of these microbes in acting as salugens and protecting against disease may explain why there is no close correlationship between regular tooth-brushing and good dental health.

Caress a baby. Feel its soft skin, and inhale that delicate, waxy scent of healthy normality on its body. Unless I am mistaken, you are smelling the microbes that live there, and which are there for a purpose.

Let no grubby little child imagine that this allows him to leave off washing altogether. It does not, of course. To feel good, and to look good, we should be clean and fresh. But if that is allowed to step out of bounds, until it becomes a license to poison and spray every member of the skin's ecological community, then we can find ourselves in difficulties. Soap-and-water, yes: vaginal and paediatric biocides, no.

The complex bacterial associations in our mouths are part of our own internal ecology, and I do not doubt that we need them. There are fairly constant populations of harmless (indeed salugenic?) streptococci in the mouth. Some of them may be assisted in their growth by their microbe neighbours: for instance, *Streptococcus mutans* is a bacterium which requires supplies of para-aminobenzoic acid if it is to reproduce. The related *Streptococcus sanguis* produces an extra supply which, at least in culture, can be used by the *mutans* species. Other streps need amylopectin supplies, and in the mouth it is likely they derive them from the *Neisseria* organisms one can find. We know how lactic acid in the mouth is believed to attack our teeth: but the *Veillonella* microbe can use lactic acid as a source of the lactates it requires for its own growth, and so help to keep acid levels down. There are some regular protozoa found in the mouth, including an amoeba (*Entamoeba gingivalis*) and the active *Trichomonas* and *Selenomonas*, and a bacterium shaped like a corkscrew known as *Treponema denticola* (a 'relative' of which is the causative organism of syphilis).

There are long thread-like bacteria found in the mouth, many of them with what seems to be a permanent attachment of small, rounded species. And the microbes that live amongst them certainly exert powerful interactions. *Streptococcus sanguis*, for example, can inhibit the growth of other streps, and it can inhibit the lactobacilli which have been accused of causing tooth decay. If you study mouth microbes, however, there is no one type which acts as an overall regulator or controller of the rest. Many of the species seem to exert different effects at different times: for some bacteria they are repressors, while for others they seem to act as growth enhancers.

So the microbiology of the mouth is exceedingly complex, and as an ecological study in its own right it offers plenty of scope for further research. One thing is certain: with a regular intake of food, and being lined (as it is) with a moist surface of dead and dying skin cells, the mouth ought to break down and decay in no time. The fact that it does not must surely be due to those salugenic microbes protecting us all, and the popular view that all organisms occupy their own ecological niche is no explanation at all. If harmful organisms are being kept in control by health-promoting species, then this is a positive action. It is no system of 'mutual overcrowding' that keeps the mouth healthy, but a selective restriction of the harmful microbes alone. Tooth brushing, though it removes food residues and is cosmetically useful, is certainly not necessary for health. Perhaps we should identify some of these protective organisms, and see if we could not culture them for use in the fight against illness. Instead of sterilizing a dressing, might it not one day be possible to inoculate it with salugens, as part of the rehabilitation of the skin for a normal existence?

Far from trying assiduously to avoid microbes, we should recognize that they constitute an unexpected and largely unrecognized defensive force: one which the unaided eye cannot appreciate, and with which the stereotyped conformity of science has yet to come to terms. We now need an unprecedented amount of research into these salugenic organisms, and when it is done we will find that those microbes which share our skin are not simply to be scrubbed off or poisoned, like so many non-fare-paying intruders, dismissed by man as a nuisance.

Quite the contrary: we need them to survive.

7 What is a Microbe?

(1) The Algae

We have done well so far to discuss so many attributes of microbes without saying much about what, exactly, they are. The microbe is a living cell; or more precisely, an organism whose body is not divided up into cells. It is useful to have a concept of what a cell looks like, if we are to appreciate it properly, and even some scientists who study cells under the microscope spend so much time looking at the toughened, sectioned remains of dead cells as to lose sight of what a fresh, living cell is like. The most accessible cell, surprising as it may seem, is a hen's egg. For all its size, an unfertilized egg is a single cell and its 'white' is of the same jelly-like consistency as the cytoplasm of the typical microscopic cell. The yolk of an egg is a single particle of stored food reserves, and in the typical cell there are many quite like it. The cell of a green plant – like an alga – has a thick cell wall (equivalent to the egg's shell) and instead of a yellow yolk body it has a green chloroplast, a structure containing chlorophyll to trap and convert solar energy. Animal cells, by contrast, are often virtually naked and either change shape to move (like an amoeba) or else they may sprout small structures by which they can swim about.

This explanation of the consistency and the 'look' of a living cell is grossly oversimplified, as the next few sections of this book will show. The variations on the basic theme are so many, and they take such inexplicably complex forms, that the fundamental concept of a cell has to be stretched and contorted to accommodate all the permutations seen in nature. But if you have had no previous idea of what a cell is like at all, then the simplistic model may well be helpful (and some of my colleagues tell me it has some relevance to the research worker who deals exclusively with stained microscope preparations, for whom the temptation to forget that the specimen was ever alive is clearly a powerful one).

The controlling centre of the microbial cell is its nucleus, a separate blob (often rounded) in which lie the genetic blueprints for the cell's structure and its code of behaviour. The microbe world shows considerable variations in nucleus structure, far more than you find in the nuclei of the millions of cells of which higher organisms are made up; in some forms there are several nuclei to cope with the day-to-day running of the cell's machinery. But in one respect the microbes are all the same. They are not divided up into cells, and this is what makes them fundamentally different from the rest of us. In some instances we find that microbe cells grow joined end-to-end, and in some of the groups (like the algae and fungi) it is easy to see how at one end of the scale we have single-celled microbes, while at the other extreme we have forms in which the single cells unite to produce a multicellular structure (seaweed, for example, or a toadstool). These have a particular philosophical interest for they remind one of the way in which microbes discovered the benefits of a communal existence, and how the many-celled higher forms of life arose.

The most important single act of living organisms on earth is the capture of energy from sunlight. It is this, and this alone, which stokes the fires of life by hoisting the Big Dipper of energy up to the top of its run. We may steep carbon dioxide and water together for an eternity but they will never make sugar; yet the self-same materials (taken one atom at a time) can be easily elaborated into the most complex of carbohydrates by a green plant cell growing in the sunlight. The harnessing of photons – discrete 'particles' of light – to drive electrons and so to reduce carbon to its energy-rich form is a form of photochemistry that happens with extraordinary rapidity, and no one yet claims fully to understand it.

We like to see this trapping of solar power as the province of the trees and flowers of the countryside, but their role is of secondary importance. The whole of the earth's flora put together, trees, flowers, microbes and all, produce 150,000 million tonnes of organic carbon every year. What is not generally realized is that 130,000 million tonnes of that are produced solely by microbes – the algae floating amongst the plankton of the sea. Each year it can be calculated that 400,000 million tonnes of oxygen are liberated into the atmosphere by the activities of the earth's green plants; but at least 350,000 million tonnes of that derive from plankton too. Far from being of academic interest, the algae which do this work are vital for the survival of life on our planet.

The algae as a whole are a varied lot. At the upper end, as it were, the spectrum extends to the seaweeds which anyone would recognize as plants, while at the other extreme lie the diminutive blue-green algae, the

Cyanophyta, which are almost certainly related to the earliest forms of green plant life that ever evolved. The macroscopic algae with which we are familiar, like the seaweeds, are not candidates for detailed consideration in a

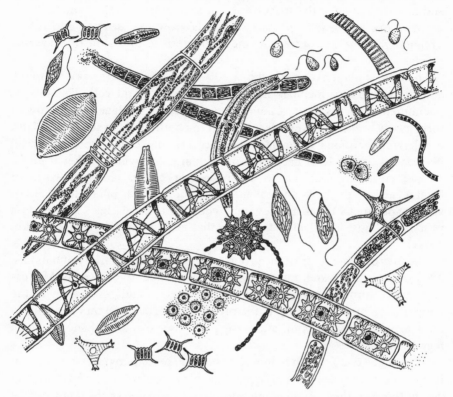

Fig 13: A selection of algae. *Spirogyra* runs diagonally across this figure, the star-shaped chloroplasts of a *Zygnema* filament showing clearly beneath it. The boat-shaped structures marked with fine transverse lines are diatoms (p. 123). Near the centre is a star-shaped group of *Pediastrum* cells, above which lies *Closterium* which looks like a boomerang and is perhaps the most familiar of the desmids. The triangular cells near the lower corners of this figure are reproductive bodies of the desmid *Staurastrum*, the vegetative form of which is the branched structure (right centre). Also present are the *Euglena* cells with their single flagellum, and free-swimming cells of the *Chlamydomonas* type which, with their paired flagella, move by means of the microbial equivalent of breast-stroke. The large filamentous alga above *Spirogyra* is *Oedogonium crassum*, in which specialized cells undertake the dividing, and accumulate successive daughter cell walls as rings around one end of the cell. Below it are the brick-shaped cells of *Microspora* (also pictured in Fig 4, p. 14). Among the other types present are the spiny groups of four small cells, *Scenedesmus,* common in standing water. For clarity, the different algae are not drawn to scale.

119

book on microbes. But, in case they seem to be strange bedfellows for the microscopic algae, it is as well if I describe them before moving on. They have no distinct root, stem, or leaf; each zone merges into the next gradually, and there are none of the special anatomical features for the conduction of sap as there are in land plants. Seaweeds grow in a strange and rather irregular manner in contrast to the characteristic precision of flowering plants.

The largest plant ever recorded, a 700-foot *Macrocystis*, was a seaweed, and the group also includes the largest recorded plant cell. This unique phenomenon occurs in the Cauperlaceae, a family of algae which grow to form mossy branches several feet long, at the bottom of the sea. Each plant, for all its conventional exterior appearance, is a single cell. There are no cell boundaries of any kind inside the individual plant, which makes it into a unique curiosity. One group of freshwater algae, the stoneworts or Characeae, are the most ancient surviving multicellular plants to be discovered. Specimens found in Devonian chert beds dating back 350 million years are almost identical to the types that grow in our streams and lakes today.

The microbe members of the group grow in a startling variety of habitats. They occur in salt water and fresh, in soil, sand, and in the oceans. Microscopic algae form extensive colonies in hot springs and geysers where the temperature may exceed 85°C for prolonged periods. Others are able to grow around freezing point and have given rise to coloured snowfields in high mountainous areas.

Some algae coexist with higher forms of life, perhaps as commensals, but in many cases they seem to play a positive role in promoting the host's health and in this way they would come into the salugenic category. They can be found in corals, adding a characteristic colour to a reef, and even under the scales of fish. *Basicladia* lives on the shells of turtles, *Characium* on mosquito larvae, and *Zoochlorella* lives between the cells of some sponges and hydroids, providing them with an extra supply of available oxygen. Some cyanophyte algae (including *Anabaeniolum* and *Oscillospira*) even exist in the human intestine, where they may play a part in the processes of digestion. They lose their colour in the dark, but will recover it if cultured in daylight and then they resume normal photosynthesis. The characteristic colour of the South American sloth is due to the growth of two algae among its hair scales: *Cyanoderma bradypodis* and *Trichophilus weckeri*. The marine alga *Prototheca zopfii* has even been discovered to break down petroleum molecules – and an oil-degrading alga could be an intriguing pointer towards a new branch of research with the control of pollution, or the production of food from waste oil and sunlight, as some of the possible consequences.

From habitats in salt springs to the human intestine, from the fur of sloths to near-boiling geysers, from high snowfields to the gnat larvae of a garden pool; the commonest plants known, and the most vital; the largest, the smallest, the longest, the most ancient and most strange plants of all – these are the algae. Some algal microbes live as crawling cells that move around like an amoeba (see p. 129) while others swim by means of fine projections from the cell wall. Most of the swimming algae are microbes, but some others – including *Volvox*, which is visible to the naked eye – grow in colonies like a microscopic golf-ball and row themselves along by the concerted action of many swimming cells in unison the tiny globes rotating like crystal spheres in the water of a lake. It is an extraordinarily beautiful sight under the microscope.

Most algae have a thick cell wall composed of cellulose, the substance most familiar to us as cellophane or cotton-wool. Cellulose is the end-product of photosynthesis; the plant makes a range of energy-rich carbohydrates including sugars and starch, and celloluse is a complex polymer containing many molecules of glucose. Cellulose does not dissolve in water, and obviously the most sensible place for a cell to store a product that will not dissolve, is outside the cell – exactly where we find the thickened cell wall so characteristic of green plants. I think this is the reason why plants developed their cellulose-reinforced cell walls in the first place, and this might explain why plant cells as a whole are characterized by this rigid, almost armour-plated boundary. All land plants feature this characteristic, and so do most aquatic forms, the only exceptions being such types as the small number of algae that produce oils or some other product instead of cellulose. Indeed one has only to glance at a microscope slide of some unnamed tissue to know whether it is of plant or animal origin – the thickened cell wall gives the game away at once. Without this feature, no land plants could exist as they do today. They gain much of their rigidity by pumping up their cells with water until they reach a state of pronounced turgor. A carrot, say, or a potato, is a rigid firm body, but if its cell walls are ruptured (by freezing, or by cooking, for example) the tissue becomes a soft, watery mass. That is what it would be like for plants in nature without rigid walls around each of their component cells, and every cell inflated hard like a motor tyre. Obviously the origins of this mechanism are important, and it is certainly possible that the insolubility of cellulose gives us the clue.

Inside the typical algal cell we find much of the space is taken up by the watery, glutinous cytoplasm; and somewhere inside this lies the nucleus, that all-important, translucent body that runs the whole machine. In algae this component is typically bun-shaped. It was once confidently predicted that the development of the electron microscope would enable us to show how the various parts of the nucleus were made up, and how it carried out its many tasks. But, apart from looking impressively larger, the nucleus has refused to give up much in the way of internal structure to the peering eye of the microscopist. Though the rest of the cell has been explored in detail, and the many mechanical features within it documented and analysed, the nucleus shows itself to be a granular mass and, so far, little else. Textbooks illustrating 'the structure of the cell' skate delicately round the problem, by devoting pages to other cell components and often passing quickly over the mysterious matter of the nucleus.

The alga most widely taught at school has long been *Spirogyra.* I am not certain why this should be: it is not the commonest type of alga, nor is it representative, and the different species have an appearance that varies considerably from one to the next. I have found students who had been taught about it in detail from textbooks and the blackboard unable to know what they were looking at, when presented with living *Spirogyra* as it occurs in nature. Reality and formal teaching are often very different things.

Spirogyra is a most entrancing, delicate and aesthetically pleasing microbe. The green chloroplast runs around the inside of the cell wall in a precise and regular spiral, giving the individual cell a spring-like appearance. This was the first alga to be written about by the most prolific of the pioneer microscopists, Anthony (Thonis) Leeuwenhoek, and its cells grow end-to-end in line, like carriages in a railway train. In this way it could be said that the filamentous algae, like *Spirogyra*, show the first attempts to grow into a many-celled plant body – yet each *Spirogyra* cell continues to function as an independent body. Except, that is, when sexual reproduction takes place. Then, separate cells growing alongside each other in adjacent filaments send out 'shoots' which meet, and fuse; and then one cell – the 'male' – passes through to meet and become one with the other, 'female' cell. The resulting spore-like body develops a thickened wall and, even when severe winter weather sets in, this capsule can survive intact, ready to produce a new filament of cells when the weather warms up again. The cells multiply by dividing in two. A cell that is about to undergo division increases in length, its nucleus divides to form two identical daughter-nuclei, and then a new cell wall grows across to separate them. Some small microbes can be seen to graze along the outside of the *Spirogyra* cells, in the manner that tick-birds hop about on a rhinoceros. Like many other algae, the filaments are covered with a thin slimy layer. As a rule the layer is difficult to study, but I have found it is possible to attach the cells to a microscope slide, and then peel the filament away from its sheath. What with this layer, and the possibility that smaller microbes help to keep the surface free of unwanted passengers, the *Spirogyra* filament seems to be well protected.

Not all algae live a sedentary life, floating aimlessly in the water of a still lake, like *Spirogyra*. Many of them swim actively, such as the *Chlamydomonas* cell which is also taught at school. It literally rows itself along with two fine, hair-like flagella that project from the cell apex. If ever an organism mastered the breast-stroke before man, it was surely *Chlamydomonas*. The nucleus lies near the centre of the cell, surrounded by a large chloroplast – but near the edge of the cell lies a structure one might not anticipate finding in an organism so diminutive: an eye. Like many swimming algae, *Chlamydomonas* is endowed with a small eye-spot, taking the form of a rounded, transparent lens with a pigment cup beneath it. Many other algae have been shown to have a clearly demarcated lens which serves to focus light on to the pigment cup – exactly like the many-celled eye of higher animals like ourselves.

The eye with which we are familiar is much easier to understand. The light receptors are the thousands of specialized cells of the retina, and they transmit the stimulus along the nerve cells to the brain – it is a mechanism that, in outline, can be grasped by a child. Indeed we are brought up to the knowledge that our bodies are composed of specialized, separate cells carrying out precise and predetermined functions. Yet in these algae, within the cytoplasm of each single microbe, is a minute structure that is also an unmistakable eye. Far from being a many-celled organ, this is a minute fraction of just *one* cell. Even the reproductive cells liberated by the humble sea lettuce that grows in green, filmy patches on our shores possess these structures. So, in coming to grips with the microbe world, we have to accept that some 'plants' can not only swim better than man, size for size, but can even see where they are going.

From the whole of the algal world, let us consider one group in a little more detail. Far from being remote from our everyday experience and unimportant, they are closely integrated into mankind's life-style. They have given us vitamin D, toothpaste, polish – even explosives; and much more besides. These are the Bacillariophyceae, to be pedantic about it; to microbiologists they are colloquially known as the *diatoms,* and that is what we shall call them here. In selecting the diatoms for consideration, one must bear in mind that they are just one group of many, and they are in this sense a small part of the community of algae as a whole. But they are singularly beautiful to study, they are economically important, and they play a vital role in the affairs of man and in the earth's ecology.

The algae specialists – phycologists – recognize some 5,000 species of diatoms, and it is likely there are as many still awaiting discovery. They occur in all types of natural waters, salt and fresh, hot and cold; and diatoms are possibly the mose widespread of all algae and the most numerous. They possess a characteristic structure which is unique and of indescribable delicacy. Each cell has a skeleton composed of silica glass, a rigid, transparent, regularly patterned framework that is repeatedly perforated with tiny apertures through which food materials and so on may pass. There are two main group of diatoms, the Centrales – which are rounded – and the Pennales, which are elongated and boat-shaped. No two diatoms are quite the same, but the patterns of marking and the shape of the outline enable the various species to be identified. Even so, no two members even of the same species are absolutely identical.

The skeleton structure of the diatoms can last forever. Silica is one of the most resistant of all compounds, and beds of fossil diatoms hundreds of feet deep and scores of miles across are found to contain these unaltered skeletal structures in exactly the same condition as that of a diatom alive today. The perforated nature of the cell skeleton makes it ideal for use as a sieve. Filters made from this diatomite rock are widely used in industry. The acids used in factories and the corrosive chemicals of heavy industry are unable to affect the silica, and highly efficient filtration can be obtained. Used in this way, the diatomite rock is simply mined, cut to shape, and fitted; no prior treatment of any kind is required.

The hardness of the silica and the porous nature of the 'shells' make them equally useful as abrasives. Minute particles of dirt are caught in the perforations, and the result is a delicately thorough cleansing process which, because of the light, fine structure, does not damage a delicate surface. Fullers' earth is a form of diatomite left by more recent diatom growth, and it is a frequent constituent of silver polish. It is also used in toothpaste.

The open, porous diatomite is very light: a cube 30 cm along the side (a cubic foot) weighs as little as 5 kilogrammes. As a consequence it has a highly absorbent nature. It is used to fill containers in which dangerous liquids are to be carried, where it has the effect of restraining the liquid so that it does not surge about. In Germany, diatomite is known as *Kieselguhr*, and the discovery in 1867 by Dr Nobel that nitroglycerine could be absorbed by the material and so handled safely, gave us dynamite. Indirectly, it also gave us the Nobel Prize.

The great stability of diatomite under extremes of temperature has given it uses as an insulator. At temperatures in excess of 1,000°F (540°C) it is more effective than magnesia or even asbestos. Diatoms from the fossil-beds have been included in compounds used for road markings, both because of their hardness and the reflective quality of the glass-like 'shell' which greatly increases visibility at night. In addition they have been used as a filler for paints and plastics – and as a bulking agent added to flour when legal controls were not as stringent as they are today. Diatomite dust is even sprinkled in the underground passages of coal mines. In this way it interferes with the spread of dust ignition, and is a valuable aid to the miner in his perpetual campaign against the risk of a self-perpetuating explosion. (British mines, which do not have such a ready supply of diatomite powder as those in the United States, utilize limestone dust as an alternative suppressor – but limestone too has microbial origins, of course).

Those are the uses of the dead, dry remains of extinct diatoms. What of the value of the living population of today? They play a vital role in gathering food materials together in the first stage of the food funnel system. Vast amounts of valuable materials – nitrates, phosphates, and the rest – run out to sea through the rivers, and would be lost to the ocean deeps were they not captured by the diatoms of the sea. In this way these 'wastes' become a staple item of diet for the diatom population of the continental shelf. As the diatoms proliferate they are gathered for food by larger organisms and so on through the food funnel system. Diatoms, for example, provide us with much of our dietary vitamin D. It is produced by these microbial cells and as they are eaten by successively larger organisms and the vitamin consequently concentrates, it reaches high proportions in fish-liver oil. Though vitamin D is formed in our skin by the action of sunlight, in practice children, particularly, need some 200 units per day in their diet. Even mother's milk on its own is deficient! It is from such oily foods that we derive the bulk of our external supplies of vitamin D – most of it produced by diatoms in the first instance.

The effect of the 'waste' chemicals in river water cannot be over-estimated either. It has been calculated that a single milligramme of elemental phosphorus can supply 800 million diatoms, which in turn provide food for a host of primary feeders (p. 93). It is well known that an increase in nutriment run-off can cause an upsurge in the growth of microbes, which in turn stimulates the growth of the organisms that feed on them. In a vitally important way, they are an essential part of the ecology of marine life, and if we managed to eliminate all 'pollutants' from the rivers we would adversely effect the balance of life around our shores. Little wonder that a great deal of mankind's waste can eventually be disposed of in this way, with consequent beneficial effects on the productivity of the sea.

Diatoms have also found a use in settling arguments over the merits of different microscope lenses, and the size of small structures, in the past. It has always been difficult to be sure about the optical quality of a lens, since so many of the features of an image are purely subjective. But the relatively consistent size of the perforations in a diatom's skeleton gave them a use as test objects. If a good lens enabled the structure to be resolved, a bad one would fail to reveal it successfully. The markings on the diatom *Amphipleura pellucida* were the most widely used test object in the late nineteenth century and have been used for this purpose ever since. The markings will only show if a high-power microscope is fitted with good lenses, and is correctly adjusted. If there are deficiencies in the system the pattern disappears. There has been some argument over the consistency of the pattern from different strains of the diatom, since the spacing of the markings may vary a little, but in normal practice this is of little consequence.

Once the size of the diatoms became known, of course, they could also be utilized as a means of comparison with other microscopic objects. Indeed the diatom was mainly responsible for the calibration of microscopy a century ago. In 1850, for example, the size of such an everyday microscopical object as the human red blood cell was variously quoted. The largest figures given were three times bigger than the actual size of the cell, and the smallest were three times too small; but reference to a given diatom pattern was able to provide some sort of universally available scale of comparison. What is more, the 'calibrated' diatom was available from any ditch, or a still pool in the country.

In life, diatoms are generally active and often move by pumping cytoplasm out through pores at the front of the cell and drawing it in at the rear, thus moving along like a tracked vehicle at speeds that enable them to cover a distance equivalent to their own length in a second. Diatoms produce oil as an energy reserve, and as much as 10 per cent or more of each cell can be composed of oil. We do not harvest diatoms in order to exploit their oil content, though I do not doubt they could be used in much the same way as such oil-forming plants as peanuts, linseed, and cotton. The culture of microbes might show us how to obtain oil from sunlight in this way.

126

Though we do not obtain oil from today's diatom populations, we are probably burning the oil left behind by the vast amounts of diatoms that were living millions of years ago. Indeed it is probable that all of our commercial oil-fields derive from the remains of ancient, extinct diatoms. Fossil-beds contain billions of cubic feet of the microbes, one-tenth of which was originally oil. In the oxygen-free environment of a huge bed of dead organisms, the stored oil would not degrade but would remain behind, percolating into hollows, rising above water layers, and gradually changing into the oil we know today. The type of oil produced by diatoms is chemically different from that found in our oil-fields, however, and so we have to ask an important question: was this natural plant oil capable of conversion into

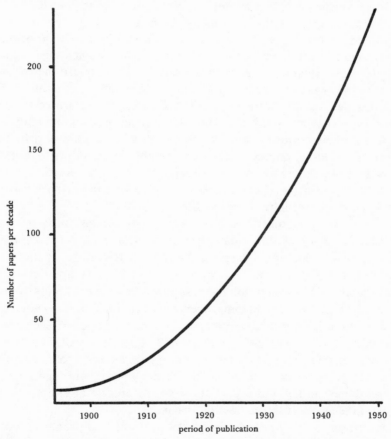

Diag 11: The growth of petroleum microbiology. The steady increase of interest in this subject dates from the turn of the century.

hydrocarbon fuel by natural processes? Research published early in the Second World War provided some strong evidence that suggested it was possible. In a mixed culture of the diatoms *Rhizosolenia* and *Nitzschia*, it was found that after the cells died, instead of the stored oils becoming oxidized or hydrolysed, they tended to combine to form more complex molecules than anticipated, and the amounts of hydrocarbon in each dead cell steadily increased. Other research workers have found in mineral oil traces of chemicals that are known to form from chlorophyll, and in some cases even derivatives apparently from the blood pigments of lower animals.

One could argue that these interesting compounds could have formed naturally, without the intervention of living organisms. But the obvious and most direct conclusion is that microbes were involved in the production of the oil – and diatoms are the most likely candidates of all.

We should not lose sight of the vast scale of operations that microbes can bring about in concert. It has been found that the plankton floating in a lake can produce 10 grammes of glucose per day in every square metre, and diatoms are known to be able to trap energy in nature at the rate of 60 kilocalories per square metre on a sunny day. A ½ hectare of lake in Wisconsin was shown to produce 1,166 kilogrammes of planktom per year, which was funnelled into the production of 132 kilogrammes of fish; and elsewhere 1,622 kilogrammes of microscopic plant life was shown to support the growth of 244 kilogrammes of aquatic animal life per hectare. So the role of the microscopic algae, diatoms above all, is of vital importance for the provision of food for higher forms of life.

Diatoms make up the diet of many other microbes, such as *Amoeba* (see p. 129). Careful study of one ciliate (see p. 134) named *Oxytrichia* has shown it can eat 90 diatoms of the genus *Navicula* in a single day. So the mass production of diatoms in nature is important to life as we know it – little wonder that the sea has been claimed to support the growth of 5,000 tonnes of planktonic algae (mostly diatoms) every year in a single square kilometre. In terms of intensive dry-land farming, such a total is inconceivable. Natural blooms of diatoms may produce as many as 2,000 million cells in a litre of water; all producing carbohydrates, all storing oil, all likely to be eaten by other larger microbes or to settle to the bottom where their dead remains form vast dunes; and all this is part of the steady process of sedimentation that goes on relentlessly throughout the oceans.

So, to this one kind of microbe, we owe a vast range of products: toothpaste and dynamite, safer transport for acids and insulation for buildings; fullers' earth, polish, and reflective road-signs; the vitamin D found in animal foodstuffs, tile-filler; water-glass, an anti-explosive agent in the mines, industrial filters, and a test-object for microscopes. They may too have given us our all-important oil-fields – which completes an amazing list of useful products from just one type of microbe.

And yet – how many people are interested in diatoms? Or have heard of them?

(2) The Protozoa

Just as the algae are the single-celled green plants, so the protozoa are their animal equivalents. These, named as the 'first animals', include the most complex of all microbes. Many of them are larger than the typical microbe cell, and some can just be seen clearly with the naked eye. What is most interesting is that the free-living protozoa can carry out most of the activities undertaken by higher animals, and they can undertake many tasks that higher creatures cannot.

The typical protozoan is best visualized as a minute droplet of raw egg: a soft, exceedingly delicate particle of translucent jelly. There is no sap-filled cavity of the kind we see in plant cells, and the cytoplasm (which in the mature cells of higher plants and in many algae merely *lines* the cell) *is* the cell. It is bounded by the thinnest membrane which in many protozoa is easily disrupted. They may be susceptible to temperature changes, pressure, and to the smallest traces of certain irritant chemicals. Protozoa observed on the microscope slide will sometimes burst asunder and gently mix with the surrounding fluid, dissolving away into the water like patterns of dust in the breeze.

Of all the protozoa, *Amoeba* is the most familiar. It is an irregular blob of cytoplasm containing a translucent nucleus near the centre of the cell. Dotted about in the jelly itself are numerous small particles and crystals, and there are small rounded bubble-like structures containing fluids and the remains of bacteria, diatoms, and other kinds of food. Some of the vacuoles contain surplus water which the cell needs to expel. They can be observed to increase in size, then move to the surface of the cell and burst just like a bubble, releasing the contained liquid. So the food vacuoles are the *Amoeba's* digestive system, and these contractile vacuoles are its 'kidneys'.

The cell moves by flowing along. One of the irregularly extended arms can be seen to act as a 'head' and the granules in the cytoplasm may be observed to flow along into it. In this way the whole of the cell gradually flows into its

129

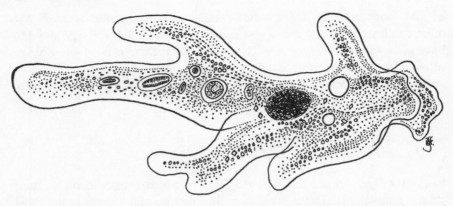

Fig 14: Amoeba proteus. The amoebae are the archetypal single-celled forms of life. For all the strangeness of some of the other microbes figured in this book, an amoeba is familiar to everyone. Its method of avoiding unpleasant stimuli, of feeding, and indeed of moving at all, are complex tasks for a single and largely undifferentiated cell. It would be more realistic to regard these organisms as posing almost insurmountable problems for science, rather than as the epitome of unrefinement and simplicity.

leading lobe, and so it moves. When it encounters a food particle, it forms its mouth at the nearest point of contact so that a dent appears in the cell membrane and the rest of the cell flows on, over and around the prey. The food is gradually digested and any indissoluble remains are expelled in the same way as surplus water – by release at the surface of the cell.

The whole effect, as we were taught as children, is like a spoonful of treacle running down a warm plate. There are no organized features within the cell, and for this reason *Amoeba* has been a standard object of study for many years. It represents, they say, life at its simplest. And amoebae are, to quote some examples from the literature over the years: 'a simple mass of organic animal matter' (1870); 'perhaps the lowest form of animal life' (1896); 'simplest member of the group' (1903); 'simplest, least organized' (1938); '*Amoeba*, one of the simplest of all animals' (1942); 'a very simple form of life' (1960); 'animalcule of the simplest structure' (1964); 'naked bit of protoplasm surrounded by only a thin plasma membrane' (1968); and so on. Two of these books quoted qualify their comments and provide more detailed accounts of these organisms, but it remains true that such allusions to the simplicity and essentially elementary nature of *Amoeba* have rubbed off on most people. It remains everyone's ideal lowly creature, the popular concept of life at its most simple and unrefined.

I think we can learn some new respect for the magnificent complexity of life when we realize that even amoebae are immensely complex, and well beyond our capacity to analyse in real, functioning terms. It is very different from that blob of treacle on a plate: *Amoeba* can move where it likes, could move uphill as readily as down, and systematically propels itself. In what way can predominantly featureless jelly undertake to contract, to flow, to undergo countless predetermined changes in viscosity, to circulate its cytoplasm, and so to move? Experiments have shown that if you watch the way the granules within the cell move, you can see that *Amoeba* has a preferred orientation, so that one part of the cell always leads the way. But how can an irregular drop of cell substance know which part of it is the 'head'? How does the cell keep itself intact, when it exists in the form of soluble complexes dissolved in water, yet living in water itself? We teach that the cell substance is composed of an endoplasmic reticulum: a network of convoluted membranes. But how can we reconcile a configuration like that with the reality of a soft and actively crawling microbe which perpetually changes its shape?

It can break its cell membrane to take food in, or to liberate wastes; it can form new vacuoles where none were before. So why does not the whole cell dissolve away into its surroundings? If it contains enzymes to digest its food, how does it not digest itself? I would answer that at once by saying that the amoebae localize specific enzymes to particular areas of the cell, so that by acting in sequence, and not all at once, they do not disrupt the cell's functioning. That is a reasonable answer, certainly: but it poses the more complex problem of how enzymes can be retained within specified regions of a homogeneous cell. These organisms are able to creep towards a favourable stimulus, temperature, or oxygen concentration. In what way can they do this? One could argue that the sense of direction is actually one of memory, so that the cell compares the amount of favourable stimulus at any given time with what was there before, and so keeps moving 'uphill' towards the target. But does this answer the question? Only if we admit that we cannot imagine how *Amoeba* could possibly aspire to the possession of memory. *Amoeba* has no component cells, and therefore no brain.

This supposedly primitive microbe can undertake many of the tasks that mankind can. It is able to move where it likes, feed, reproduce, respond to stimuli, and so forth. But it has some useful advantages over man, for it can build a protective capsule around itself if its environment dries up, and it regulates its rate of reproduction to match the food supply. Like the typical microbe, *Amoeba* multiplies by dividing in two. The cell grows to double its original size and then takes half an hour or so to undergo fission, before the resulting daughter-cells go their own way to repeat the process. There is nothing dead left over. Mankind gives but a single, insignificant cell to the next generation, and he dies in the fullness of time. But microbes like *Amoeba* live on, their entire, immortal bodies splitting into twinned descendants at predictable intervals. The human population doubles in size every 25 years, and is causing concern as a result. These microbes *can* double their numbers nearer a million times as fast, but for them proliferation is a controlled and self-limiting phenomenon. That is something else we have yet to imitate ourselves.

Finally, what of the cliché about the shapelessness of amoebae? It is often said that '*Amoeba* can take up any shape', but it is not so simple. No two cells are ever the same shape, that is true; and the outline of each active *Amoeba* changes perpetually. The chances of two individual cells having an identical appearance under those conditions are the same as the chances that you would make two ink-blobs, pressed between paper, that were the same; or that you would drop two identical drops of water on to concrete and find that they made indistinguishable splash shapes. But none the less, for all the variation, you can still tell one species from another. *Amoeba proteus*, for example, has longer projections than *Amoeba discoides*. Microscopists who study amoebae enough become able to tell at a glance which species are which. What the brain is doing is analyzing variable factors which we have still to identify – ratios between length and width of the cell's projections, size and distribution of particles within the cell, something about the degree of roundedness of the outline, and so on. There are criteria which enable us to tell one from another, and they are of the same vague kind as the distinguishing characteristics that tell one, at a distance, the difference between a youth and a man wearing the same clothes, or which reveal hints of hidden emotions in the human face. The criteria by which visual distinctions are made have never been adequately investigated, and the way microbiologists distinguish between organisms without being able to say quite what it is that enables them to, is one pointer to a new form of understanding that could explain many aspects of decision making and judgement in the broader area of human existence.

The demonstration of protozoa swimming actively about under the microscope was a popular ruse of the pioneer microbiologists of the nineteenth century. As a party trick, it was very simple. You took a pinch of wheat, a little chopped straw, or some other dry vegetable material. This was added to a vessel of boiling water, which was then left to cool and to stand for a week or so. At the end of that time it was ready for examination, and a drop placed on a microscope slide showed an astonishing assortment of moving organisms which could be seen to glide, to swim about, to cavort, or to whirl like a Catherine-wheel. Since they were always found in these infusions the microbes were given a collective name – the Infusoria. Later it became clear that this term was too vague for scientific purposes. We are more interested in a name that describes what the organisms are, rather than merely where they came from. It turned out that most of the Infusoria were microbes covered with fine, hair-like, flickering projections like a covering of moving fur. Ripples of movement passing across these, like wheat in a wind-swept field, served to propel the cells along by a kind of frantic rowing. The beating threads were christened cilia (from the Latin *cilium* = eyelash) and the organisms themselves became known as ciliates.

Where did they come from? If they occurred in water that had been boiled and left to stand, they must have arisen from resistant spores. Indeed, some ciliates, notably *Prorodon griseus,* are better known as cysts: the encysted form is more common in nature than the free-swimming organism. But this is not the satisfying answer one would hope for, since many ciliates have never been observed in the cyst form. We are sure they must produce cysts, or they would be unable to get into flasks of infusion as they do; but in spite of prolonged and detailed investigation they have never been seen. The fact that this is so of ciliates like *Paramecium* and the giant *Spirostomum** – – which have been studied for well over a century – is a remarkable example of the limits to our knowledge about the protozoa.

The ciliates typically feed by grazing on bacteria, which they sweep up into a funnel-shaped gullet set in the cell wall. A flickering curtain of cilia directs a steady current of water through this system, and food particles can be absorbed by the cytoplasm at the far end. The particles move in a predictable pathway through the cell, are eventually ejected through an anal pore at the surface. Like bristles on a brush, the cilia themselves take up a regular pattern on the ciliate cell wall. Their movements and synchronization are controlled by a network of fine fibrils that lie beneath the sculptured pellicle covering the cell, like the wiring of an electronic communications system. It is tempting to see these as a kind of nervous system, until we remember that a nervous system is composed of nerve cells in higher, multicellular animals and here we are talking about something within a single cell. Nothing that we could call a nervous system could – by definition – exist inside a protozoan.

To sit and watch ciliates at their everyday work is an endless fascination.

---

* Shortly after writing this chapter I believe we were able to observe for the first time a *Spirostomum* forming cysts. If this is indeed the case, then that is one microbe off the list.

Stentor, one of the largest of them all, looks like a small, lilac-hued comma to the naked eye. Under the microscope it is revealed as a fantastic trumpet, glistening and quivering with life, a halo of cilia around its open end funnelling food into its gullet. Sometimes it will break free and glide like some weird submarine in a fable to find a new point of anchorage. On the tentacles of the pond hydra you may see *Trichodina* running up and down like a pulley or a train-wheel on its track; and then there is *Lacrymaria*, shaped like an elongated tear-drop, which hunts its diminutive prey looking for all the world like a stylized Loch Ness Monster.

Paramecium is perhaps the most familiar ciliate, since it is the type studied widely at school level. Why this should be is enigmatic: it is an exceptional organism in many ways and, like *Spirogyra* (discussed in the last chapter), it is not the commonest of the group. Its structure is hard to make out, which does not help the student who is trying to become acquainted with the protozoa, and I do not doubt that something like *Vorticella*, with its unmistakable structure and entrancing behaviour, would be more interesting to study and easier to culture, too. The permanent microscope preparations of *Paramecium* which are sold for teaching purposes show a distorted and contracted version of the living cell and – even when these inevitable artefacts are kept to a minimum – it is difficult to gain from them a feeling for the living microbes.

Paramecium moves in a steady spiral, collecting food in an oral groove that runs across the cell, running backwards and moving from left to right. You would expect that a slanting groove like this would make the cell rotate from right to left as it moves forward. In fact it inexplicably moves in the opposite direction. In case that sounds like a necessary adaptation, there is one species – *Paramecium calkinski* – which does rotate from right to left. But all the species, including this one, rotate to the right when they swim backwards, which upsets the otherwise sensible mechanical answer to the problem. Why this should be is obscure.

Paramecium swims at a leisurely pace, collecting bacteria and occasional algae as it proceeds, though when in a hurry it can cover ten times its own length in one second (travelling, that is, about 2 or 3 millimetres). If faced with an obstacle or an unpleasant stimulus it reverses the direction of its beat, swims backwards, and then starts off again in a different direction. This phenomenon is well known, and it makes *Paramecium* seem simple and mechanistic in its response to a stimulus. But what happens if we surround the microbe with an unpleasant stimulus – a bright light, for example? Then it does not merely reverse its ciliary beat, but it twists and turns, hunting for a dark corner. Its movements are every bit as coordinated and refined as those of a higher animal, an insect, say, in such a situation. A system in which the direction of propulsion reverses on contact with an unpleasant new environment is easy to analogize to a mechanical model. But the frantic avoiding movements made by *Paramecium* in my all-round stimulus experiment is inconceivably complex by comparison. *Spirostomum* is larger, over a millimetre long, and it is clearly visible to the naked eye, looking like a minute speck of cotton. It tends to graze, too, though one can find it hunting about among dead leaves and decaying matter at the bottom of a shady pond. I have filmed it nosing about among obstacles to its progress, gently, and with a kind of deliberate precision, searching about like an elephant feeling with its trunk. This is no simplistic, primitive cell; the movements are coordinated and deliberate. It is one cell doing what, in the normal way, one could only imagine billions of cells undertaking.

An example of a ciliate which has carried this form of structural adaptation as far as it can is *Epidinium ecaudatum*. It is one of the ciliates found in the rumen of sheep and cattle, where it feeds on bacteria that degrade the animal's grassy diet, and it is one of the forms of microbe protein that they eventually digest. In its normal environment, *Epidinium* can travel as much as twenty times its own length in a second (equivalent to a boy swimming faster than a power-boat), but as it cools on the microscope slide it becomes more sluggish, and is easier to observe.

At the leading end of the cell is a clearly marked mouth opening, through which the bacteria which *Epidinium* hunts are taken into its cytoplasm. Nearby, just where one would expect to find a brain in a higher animal, there lies an unmistakable, brain-like solid concentration of protein. From it, fine branches run to encircle the gullet and other tenuous threads run from this central body to connect it with the cilia which propel the organism along. Let me repeat: this cannot be the equivalent of a brain in any sense in which we understand the term. Brains are collections of nerve cells, communicating through an intricate network of branches sent out between one cell and the next. In *Epidinium* there are no cells. But it seems certain that a whole range of activities are controlled by that little protein organelle, the motorium, as it is called – and this could be a good reason for us to reconsider the way brain cells function in man. It is accepted that nerve cells funtion in an 'on' or 'off' mode – 'go' or 'no-go' is the equivalent term used in electronics – like the units of a computer. But might this primitive concept be mistaken?

If a single cell like *Epidinium* can have complex functions going on within itself, then it is at least possible that nerve cells in man could function in a

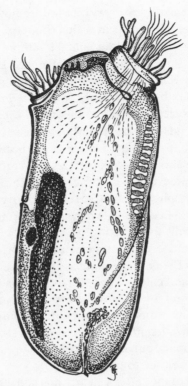

Fig 15: An *Epidinium* cell. Many of the structures of higher animals seem to be imitated in this structurally complex microbe. Between the two groups of cilia lies a brain-like motorium, whilst running down the cell is a segmented skeletal plate reminiscent of a vertebral column. Mid-way along the opposite side is the kidney-like contractile vacuole which discharges through a permanent opening like the 'anal pore' at the lower extremity. The darker bodies are the two nuclei.

manner far more complex than the simple go or no-go principle suggests. Perhaps a single brain cell can carry on some of the integrative processes of thought and memory within its own structure (that is, intraneuronal, as opposed to interneuronal activity). Brain models have tended to become firmly based on the go or no-go principle, but if that is all neurones do then they would have to be the most primitive and unenterprising cells in our bodies. In reality they are nothing of the sort. They are so highly specialized that they have lost even the ability to multiply, so that a person loses brain cells steadily throughout life, and they are never replaced. Additionally, we know that the resting nerve cells of the brain consume relatively large amounts of energy foodstuffs, which is a second pointer to highly specialized activities taking place. Perhaps brain cells function in a more complex manner than we have suspected in the past: and it is interesting that one model for how this could be arises from a consideration of how microbes behave.

In *Epidinium* there are two sets of cilia which are developed into specialized finger-like projections, the larger of the two sets being arrayed around the mouth aperture, where they wave about and direct the food into the cell. Undigested remains are eventually expelled through an anal pore. This feature is fairly typical of the ciliates, and so is the presence of two nuclei in each microbe. The larger of them, known as the meganucleus, controls the routine functions of the cell's machinery. In *Epidinium* it is shaped like an elongated lozenge or a flattened banana. Nestling into it, like a tennis ball on a pillow, lies the smaller micronucleus which contains the genetic blueprint for the cell. In the ciliates which have a phase of sexual reproduction it is only the micronucleus that is involved. The two cells ('male' and 'female') swim alongside and fuse. The micronucleus in each often undergoes some complex divisions and then one or more of the resulting nuclei are exchanged before the two parent cells part and undergo further divisions themselves. In some ciliates several different forms of sexual reproduction are known, and the entire range adopted by microbes in general would certainly to justice to the *Kama Sutra*.

Running along the interior of the *Epidinium* cell facing the nuclei is another extraordinary structure. It is a kind of backbone, a skeletal plate of stored food reserves, and it is sculpted in a way reminiscent of the vertebral column in higher animals. This skeletal plate is composed of polysaccharide (i.e. starch-like) materials, which can be used as a food store for the organism, should the need arise. There is no functional similarity between this skeletal plate and a backbone, of course; the calcium salts in bone are not an energy reserve, whereas the polysaccharides stored by the microbe are; and the development of the two systems is quite distinct. But none the less, there is a kind of 'skeleton' in this diminutive microbe, no more than one-tenth of a millimetre from end to end.

So here we have microscopic forms of life which, within one cell, have organelles of movement, mouths, muscles, even sometimes a skeleton of a sort; they may have a kind of nervous system, perhaps a 'brain'; and their relatives can see where they are going through eyes that are clearly recognizable as being built on the same pattern as the human eye. Yet all this cannot be explained away in terms that we apply to higher life forms. Eyes, brains, and muscles are the structures developed by cells that work in concert. We are now dealing with strange, unfamiliar, non-anthropomorphic forms of life, and superficial comparisons with many-celled animals will easily get us into difficulties. On the other hand, exactly what interpretation we ought to use in their place is none to clear either.

In structural terms, these protozoa are the most sophisticated microbes of all, and they offer vastly intriguing objects for study. Yet they have been widely ignored for the last half-century. Much of the detailed work of observation and documentation was carried out in the late 1800s. The overwhelming majority of microbiology texts do not deal with them in any detail, if at all; and even the best devote only a small part of the discussion to protozoa. It would be invidious to cite a particular example, but one of the most succinct and erudite works on microbiology published in recent years, which I have just looked through, devotes less than four pages of the text, and just four illustrations, to the entire subject of the protozoa; and that out of a total of over 750 pages! It has been stated that at the time of her death in 1956, Professor Doris Mackinnon was the only protozoologist who taught undergraduates in the whole of Britain. Even today, it is most unusual to find a protozoologist who was trained by protozoologists, so that many of the specialists acquire their knowledge through practical work with the organisms after training in a different (if related) subject. That is no bad way to do it – but it does mean that the way some microbiologists gain their education is much the same as one might have found several centuries ago. In many ways the study of the protozoa remains a Cinderella science – a mysterious corner of the microbe world which is perhaps still waiting for its prince.

(3) The Fungi

Though the term *fungus* is familiar to biologists and non-scientists alike, the toadstools and mushrooms are far from being representative examples of the group as a whole. It is easy to see a patch of toadstools growing in woodland and to assume that they are constructed along the lines of green plants, with roots, stems, and so on. If you pick a toadstool, you can see the fine white fibrils which run from its base into the soil. These are the hyphae, a series of tubular cell-like structures that extend for perhaps metres through the soil. The fungus that we see growing above ground level is only the fruiting body, where the hyphae grow in a solid mass so that the spore-forming areas are held erect in the breeze, where the spores have the best chance of distribution. The real fungus colony is the mass of hyphae that a cursory examination does not reveal. Often these fungi grow in association with the roots of trees and other forest plants, and the fruiting bodies are indicators of the vast mycorrhizal system (p. 44) which is giving the green plants their supply of recycled nutriment.

Toadstools are not the frightening, distaseful, dangerous things we were brought up to imagine. It is unlikely that most people will have read the words above – 'pick a toadstool' – without a slight shudder, for we have long been taught that they are deadly. In reality, even this belief is wrong. Toadstools are, in the main, harmless. A considerable proportion of them are edible, and some are extremely tasty. The poisonous fungi are very much in a minority, but they are dangerous because in some ways they mimic well-known edible species and have, in the past, been eaten by mistake. In quantity they could be fatal, but even the most toxic of toadstools could be picked and handled with perfect safety. The old tale that even to touch a toadstool, or to lick one's fingers after doing so, could kill, is nothing more than superstition. Such commonplace, familiar garden plants as the Christmas Rose (*Helleborus niger*), the Monk's-hood (*Aconitum*), and the delphinium plant, not to mention the laburnum tree, are more dangerous than the poisonous fungi. We cultivate castor seeds as a pot plant, yet two seeds could kill a child of five. While we accept such traditional homely species, it is entirely unjustified to show such an hysterical dislike of the relatively innocuous toadstool.

There is some historical justification for suspicion, however. Several toadstools produce hallucinogenic compounds, so that poisoning by them produces bizarre and flamboyant behaviour. We can cite mushroom-worship by some of the indigenous American cultures as one example of this, and the caterpillar's mushroom which Alice found to alter her size so conveniently is very likely another. Some of these fungus fruiting bodies have a striking appearance (notably *Phallus*, which looks like a ghostly penis standing erect in the woods), and that has also contributed to the legend. Others suddenly appear within a matter of hours, far faster than normal green plants can grow. This is an example of their ability to reproduce at the rates we expect of colonies of microbes, and is one way of producing food for human or animal consumption on a large scale. Freshly harvested mushrooms contain 90 per cent water, and the dry residue is itself 40 per cent protein. The food value of fungus protein is higher than that of typical vegetables, though not quite that of meat, and the characteristic odour and taste of mushrooms seems to be derived from the spores. The early harvesting of immature fruiting bodies, therefore, might give us a source of rapidly produced neutral-tasting protein food. The present annual production of commercial mushrooms totals 250,000 tonnes, of which roughly 10,000 tonnes is dry protein. The British Glasshouse Crops Research Institute has

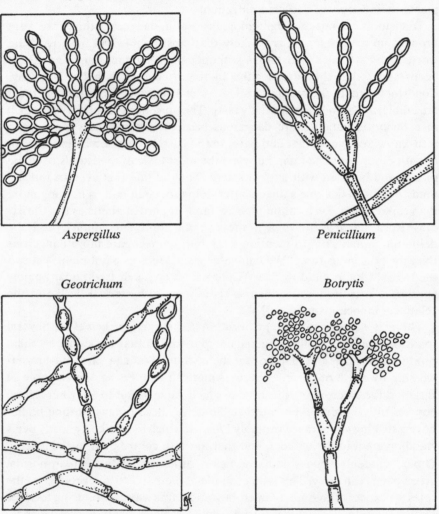

Aspergillus

Penicillium

Geotrichum

Botrytis

Fig 16: Four microscopic fungi. The spore formation systems of fungal microbes vary considerably. *Penicillium* is familiar, and is named from its appearance, which is similar to a painting brush (Latin: penicillus). In *Geotrichum*, by contrast, the spores form within the hyphae themselves. Of the types shown, some commercial uses are already known for *Penicillium*, *Aspergillus* and *Botrytis*.

calculated that over 60,000 kilogrammes of dry protein per hectare could be produced in a single year, if mushroom farming were expanded. Even that is one thousand times as much food as cattle rearing would provide.

But the familiar toadstools and mushrooms are in the minority in the fungus world, for most do not aspire to anything so large but grow in small colonies. The most the unaided eye is likely to see of them is a trace of mildew on a slice of bread. But this is the world of the microbe we can see – and most fungi are microbes in every sense. The mildewed patch that catches the attention is at first glance no more than a small patch of stale-smelling powdery, grey dust. But the hand lens or microscope reveal it for what it is: a miraculously delicate, spun-silk structure that is gracefully proportioned, and meticulously constructed out of separate hyphae.

Many fungi are more diminutive even than the mildews. Yeast looks *en masse* like nothing more exciting than putty, and it is the microscope which allows us to see what yeast is really like. It grows as egg-shaped cells which reproduce in an unusual manner, for they do not divide in half. Instead, each growing yeast cell produces a small protrusion on its surface which gradually increases in size until it is almost (but not quite) as large as the parent cell. It then may produce a protrusion of its own, and so repeat the process. Not until the little group of cells numbers perhaps five or six does it break up, each portion of the group continuing the process. This method of reproduction is aptly known as budding, and it is occasionally found in other groups of microbes. Some bacteria and protozoa feature a kind of budding – but, to redress the balance, there is a group of yeasts (the *Schizosaccharomyces*) which do not bud at all, but which divide in half like normal cells.

Some other fungi exist as separate, free-living cells which crawl about like amoebae. If you saw them through a microscope, then you would think they were amoeboid protozoa. It is an entirely reasonable diagnosis, at least until they show their incredible method of spore formation. The best-known member of this group of slime moulds is *Dictyostelium*, and for its free-living phase it crawls like any amoeba, hunting around in the wet, dead leaves of woodland for bacteria and other smaller microbes on which to feed. In culture, its cells look like small, undistinguished amoebae with nothing to suggest that they could undertake an astonishing transformation. But if we film the cells at time-lapse speeds (so that the projected image is greatly speeded up) we can see something unexpected beginning to occur. As the colony ages, and the time approaches for spore formation, the cells start sending out signals. They emit small amounts of chemical messengers, which the surrounding cells respond to. A film of a culture containing hundreds or thousands of the cells shows that the 'signallers' seem to beam out their message, and the result shows in a time-lapse film like ripples spreading out from a pebble dropping into a still pond. Some people say it looks like an animated cartoon film showing how radio transmitters work. For several hours occasional cells send out these messages, until suddenly the behaviour of the whole community changes.

They stop feeding and crawling around aimlessly, and begin to home in on a central target cell. The appearance changes to something more like an irregular star-like, splattered mass of life, for all the amoebae join together and start to head in towards the mutually agreed centre. Occasional cells that one sees do keep crawling on, and one can watch them move right through one of these radiating arms of the colony, passing through the hurrying mass and out the other side, still moving on as though nothing had happened. Whether they belong to a different species, or are simply 'deaf' to the message, is not clear.

Eventually the mass of cells produces a rounded body, shaped like a minute slug. Now the cells begin to function in a coordinated way, for this slug-like 'creature' develops a tapered head which begins to nose its way

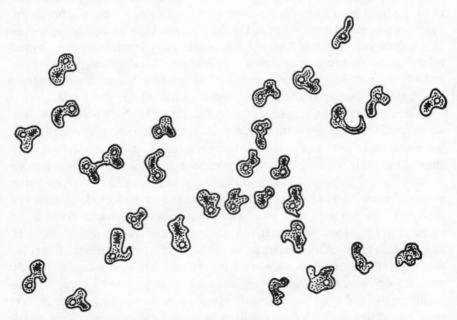

Fig 17 (a): In the free-living state, the cells of *Dictyostelium* exist like amoebae, and crawl independently through the decaying leaves of moist woodland soil.

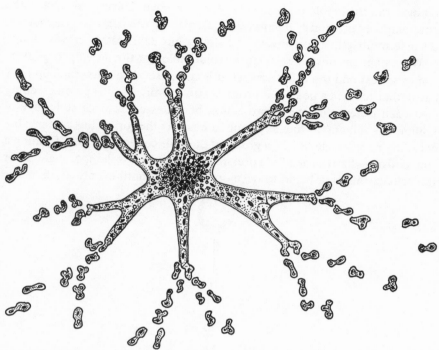

(b): On receipt of a chemical signal sent out by a central cell, the independent amoebae begin to crawl together producing a star-shaped mass which gradually contracts into a single body.

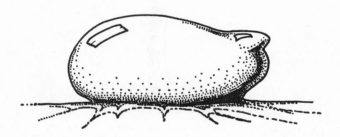

(c): The slug-like rounded body of cells now begins to function like a multicellular organism, and crawls purposely along to find a suitable site for spore formation.

143

along – and the whole body of cells starts to move. During this phase, the community of previously unconnected, free-living cells acts as though it was a single multicellular organism, and slowly but deliberately crawls along.

What happens next is even more remarkable, for the moving 'organism' comes to rest and the cells of which it is composed begin to crawl up over each other to form a peak that projects into the air. As they go on creeping past each other, this vague projection becomes a tower of cells several millimetres tall and as fine as a human hair. At the top of this remarkably coordinated pinnacle of life, a group of the cells collect together, forming a rounded mass like the head of a dressmaker's pin. This is the sporangium, for the cells in this area begin to develop hard walls until eventually they are

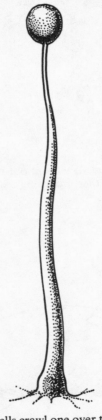

(d): After coming to rest, the cells crawl one over the other to produce a tapering, hair-like spore-bearing stalk (c.f. *Rhizopus*, p. 23). The cells at the apex develop resistant walls, and are dispersed by wind currents in the form of spores. In this stage the fungal attributes of the organism are clear.

released, and blow away like any other fungus spores.

There are several mysteries involved in the elaborate procedure, and some of the answers that have been put forward are not really solutions to the problem at all. The chemical cyclic AMP is known to induce the migration of the separate amoeboid cells, for instance, yet this tells us nothing about the way in which they detect it, or the direction it is coming from; and it leaves unanswered the complex questions one could pose about the nature of the amoeboid movement itself.

More intriguing still are the philosophical difficulties of separate cells becoming a coordinated whole, and how in the sporing phase some of the community sacrifices itself in order to allow some other cells to escape, as spores. In the field of zoology, the concept of altruism is becoming more widely accepted as examples of 'sacrifice' are discovered. The drones that die, the workers that are mortally wounded when they protect their bee community from predators, are examples of altruistic behaviour in the insect world; and birds that risk their own lives to protect their young come into a similar category. But here we have independent, free-living cells forming a coordinated community structure in which the favoured few are liberated as spores, while the majority go into making up the fine pillar support on which the spores form. This is an example of microbial altruism – something quite unheard of, and difficult to reconcile with the notion of independent microbes.

The spores I have mentioned so far have been characterized by the possession of a thickened, resistant cell wall. This enables them to survive dispersal by the wind, and the remarkable behaviour of the *Dictyostelium* fungus is in functional terms merely a means of getting airborne. But not all fungus spores are carried by a breeze: a great many fungi are aquatic, and are adapted to spread themselves through water or through wet soil. There is no virtue here in a resistant spore wall. Quite the reverse: these spores need to move. Many of them travel about by amoeboid motion, but the bulk of the spores produced by these microscopic fungi swim by means of flagella in a manner similar to that of the motile algae (such as *Chlamydomonas*, described in Chapter 7). In these types, the normal fungus grows as fine threads of hyphae, and when the spores are produced they form in the terminal region of a hypha. A split, or a small rounded opening, appears in the end of the sporangium and the swimming spores emerge. Often they swim away before losing their flagella and commencing growth into a new colony. But in many species sexual reproduction takes place during this phase. When this occurs, two of the free-swimming spore cells meet head on in a kind of microbial embrace. Eventually they fuse, and from the resulting cell the life cycle continues anew. Sexual reproduction in the fungi takes many forms, and in some cases the organism goes through a series of different forms before it arrives back where it started. We used to think of each different phase as a distinct species, and it was only the most painstaking research that eventually revealed the truth of the matter.

It is often said that fungi are organisms of decay. This is perhaps misleading, taken at face value, for all organisms break down foodstuffs – and therefore 'cause decay' – during their normal processes of metabolism. In the sense that we eat our meals, but excrete urine and microbe-rich faeces, mankind could be said to be an 'organism of decay' too. If we restrict the term to mean the breakdown of organic matter found in the soil and elsewhere, it is still unsound, for there are far more bacteria involved with that process than there are fungi.

What is true, however, is that fungi are able to utilize as food materials, substances that have only a little chemical energy stored in them. They are often found in low-grade materials after other organisms have finished with them. So it is with cheese, for instance: the available energy stored in milk is partly used by the bacteria which convert it to young cheese, and it is then that fungi are able to extract further energy from the residue – even when most other microbes might have given up the prospect. Such fungi are typical of the blue-veined cheeses, the colour being the result of their characteristically tinted spores. Such a cheese represents a protein-rich food at the lower reaches of the energy scale, when much of the carbohydrate content has been converted to microbe protein.

Some fungi coexist with other types of living organisms. I have referred to their intimate involvement with forest trees, and the way they may exist within the living cells of many higher plants. In many cases they act as salugens, by positively promoting the healthy existence of the host. Yeasts that are found on the human skin may act in the same way, and it is possible that the yeasts which habitually grow on the skin of fruits protect them in some way. It is these naturally abundant yeasts on fruit skin which doubtless gave rise to wine, when mankind first extracted and tried to store the sugar-rich juices.

We ought to try new ways to culturing fungi as an aid to our civilization, and this is a lesson that has already been learned by some species of insect. Surprisingly, they deliberately raise colonies of fungi for their own purposes. Organic matter (such as leaf fragments, in the case of the leaf-cutting ants *Atta* and *Acromyrmex*) are carried by the host insects into underground

146

chambers where the fungus grows on the remains and is tended by the ants. Only on occasions do the insects eat the fungi; it seems they supply an air-conditioning service for the ants, both by helping to maintain high humidity, and by helping to keep the nest warm. The heat generated by the life processes of the fungus serves to warm the air underground – an important factor for cold-blooded creatures such as ants, which become dormant and inactive if the temperature falls.

The scale insect *Septobasidium* and the Ambrosia beetles live in association with certain specific fungi. Indeed in some cases there is just one species of fungus coexisting with a given insect, and the relationship has gone on for such a prolonged period that the two do not exist separately. This is symbiosis (literally: 'joint living') in the true sense of the term, for the two organisms rely on each other if they are to survive.

In function the fungi play an important part in maintaining life, and in form they are remarkably diverse – often beautiful and mysterious organisms to study. To lump them together as primitive agents of decay is not only shortsighted, but it prevents us from tapping a powerful source of industrial energy – and prevents us appreciating the vital importance of fungi to us all.

Incidentally, there have been many examples of the 'record-setting' abilities of microbes as a subsidiary theme of this book, and fungi have exceeded all the great bank robberies and art thefts in establishing a world record for criminal misappropriation. In the mid-1960s it was reported that Italian industrial espionage agents stole cultures of drug-producing fungi from the American Cyanamid Company. The data and cultures were worth $24 million (£8½ million) at that time, far more than the value of anything else that has ever been stolen in history.

(4) Bacteria and Viruses

Here we seem to be on familiar ground, for as we have seen it is the word 'bacteria' that most people associate with microbes. But though the word is familiar enough, the organisms themselves are not. For all their importance to us, the appearance of bacteria is not in the least familiar and in many reference books one can see drawings of bacteria that give little or no real impression of what they are actually like. The greater publicity which bacteria have attracted over the years has served only to give them a larger platform for misrepresentation, I fear.

There are three principle types of bacterial shape: spheres, rods, and spirals from which have come the latinized descriptions – cocci, bacilli, and sprilla. Some of them exist in one shape for part of their growth phase, and as another shape for a later phase, a phenomenon known as pleomorphism. Bacteria are small. They are smaller than most of the cells we have described so far: if one could talk about a typical cell, then a typical bacterium is ten times smaller than that. Bacteria are smaller than the average cell nucleus, and because of their small size, they are not so very far from the resolution limits of the light microscope and reveal little in the way of internal structure. Even the electron microscope has disclosed little of the complex working of bacteria, since their machinery is so compact that the organized structures we recognize easily enough in larger microbes are often absent.

One interesting link between bacteria and fungi seems to be the way that some of the rod-shaped bacteria form spores. It is not that they form a protective cyst wall around themselves (as we see in the protozoa) for the spore develops within the bacterial cell, and is later released when the remains of the cell disintegrate around it. The electron microscope shows that some of these spores have a patterned, armour-plated outer layer. They are highly resistant: many can withstand prolonged periods of boiling, and

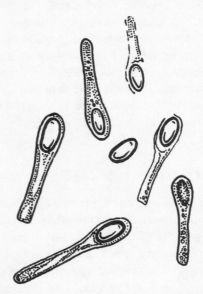

Fig 18:   Some of the most resistant spores of all are formed by *Clostridium* bacteria. They are possessed of a reinforced cell wall, and are released when the surrounding parent bacterium breaks down.

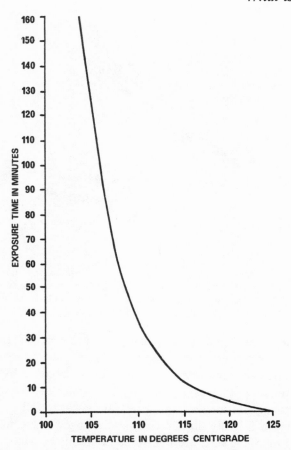

Diag 12: Death curve of spores similar to those shown in *Fig 18*. Note that the higher the temperature, the shorter the survival time of the spores. The graph reminds us that spores will survive for several minutes at temperatures well above boiling point.

some can be raised to higher temperatures and still germinate afterwards. They can survive in a totally dry environment for prolonged periods, or be frozen with impunity. We are not sure how they manage to be so resilient, and their behaviour sometimes seems to defy many of the basic rules about the nature of living organisms. The components of heat-resistant bacteria, for instance, are easily denatured by heat when they have been extracted from the cell. By contrast, the same materials left in the cell's protective matrix are a thousand times more resistant to boiling. Clearly there is some powerful force at work to keep them intact, but no one is certain what it is.

Some bacteria swim actively by means of flagella which are too fine to be

149

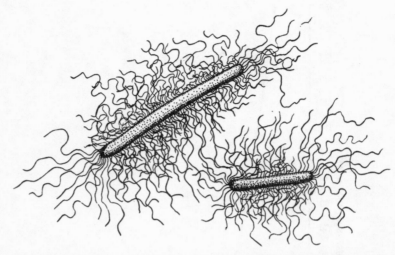

Fig 19:   Bacterial flagella. In some types, as in this *Proteus*, the mass of flagella seems so complex and overcrowded that it is difficult to see how coordinated movement could be obtained. But obtained it is: many bacteria of this kind can outpace an olympic swimmer many times over, size for size.

observed, in the normal way, by an optical microscope. Special stains can be used to thicken them, so that they can show up in slides prepared for the purpose, and one can demonstrate them on living bacteria by the use of dark-ground microscopy. This is a technique in which the flagella and the bacteria are strongly illuminated, against a black background. In principle it is like observing a brightly illuminated satellite against the night sky, as clearly visible as moving star, while it would be totally invisible in daylight. Dark-ground microscopes are infrequently used today, but they are invaluable for the study of very fine structures; and the way that flagella propel bacteria along is one of the outstanding puzzles that the technique may yet solve.

A single bacterium, alighting on a suitable substrate, multiplies by division at a rate that is normally up to once every twenty minutes. After a few days the resulting microbes form a rounded colony of bacteria which is clearly visible to the unaided eye. The swimming forms spread across their growth medium and produce a flattened and irregular growth, while the surface appearance of normal colonies (they can be rough, smooth, shiny, transparent, and so on) can be an important diagnostic criterion when it comes to the identification of the organisms.

Some bacteria have a well-developed sense of direction, or so it appears.

150

As their cells divide, they do so in a slightly asymmetrical fashion and as a result the colonies that form develop projecting arms across the medium, and the arms always turn in the same direction (i.e. they curve either to the left or right). Under the low-power microscope, the colonies have the appearance of a wig of curled hair. Such precise conformity is unusual, and it produces an attractive growth. Perhaps more remarkable are the motile, swimming bacteria of the genus *Bacillus* which orientate themselves in the same direction (like a school of fish) and swim in concert. Instead of

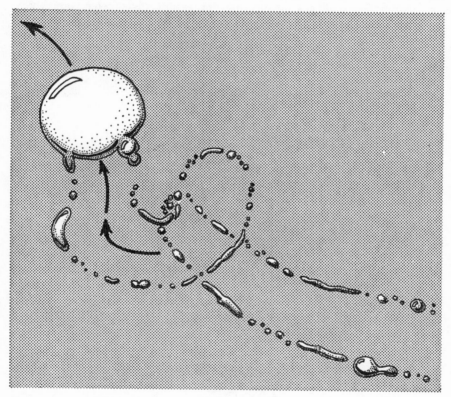

Fig 20: A crawling colony. A few bacterial types are able to move in unison, showing a corporate sense of direction. When this occurs the entire colony, which may be a couple of millimetres in diameter, moves slowly along and leaves behind a trail of smaller, daughter colonies.

spreading across the medium, as most motile bacteria do, the entire colony of these organisms moves slowly along, leaving behind a clear trail of smaller, daughter-colonies.

This ability to act in concert is carried to a further extreme by the *Chondromyces* bacteria which parallel the life history of *Dictyostelium* (see p. 142). Normally independent, the individual cells have the ability to congregate at one site. They then flow up around each other, producing a raised colony, and in time reach out to form a branching structure from the rounded branches of which some of the individual microbes are eventually shed as spores. The mature fruiting body looks for all the world like nothing more than a fairy-tale lollipop tree. Might we not learn something about the way mammalian embryos develop by unravelling the mechanisms which cause separate bacteria to aggregate and specialize in this way?

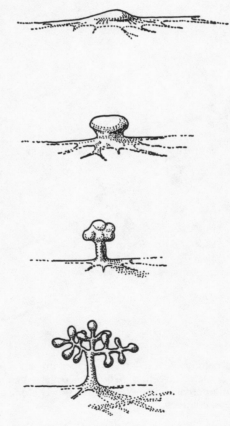

Fig 21: A microbe 'lollipop tree'. Bacteria of the *Chondromyces* type congregate in the manner of *Dictyostelium* cells (p. 142). The entire mass of independent microbes then crawls up over itself to produce the characteristically branched fruiting body, which is visible to the naked eye and looks a little like a small, soft, moist feather.

To observe the consistency and appearance of bacteria *en masse,* draw a toothpick or the point of a pin lightly down the crevice between two teeth and examine what is removed. The matter that comes away is creamy in colour and moist to the touch. Gently rubbed between the fingers it has a soft, sticky consistency like any other typical mass of cellular material. Apart from traces of food, this is a portion of microbe growth containing several hundred thousand organisms. Their presence in the mouth is not harmful, indeed they may well turn out to be an important preserver of health (see Chapter 6).

Bacteria are typically cultured on a gel made with agar, a seaweed extract, instead of gelatine. A prerequisite for the culture of many bacteria is a temperature around 37°C (body heat), and at lower temperatures than this

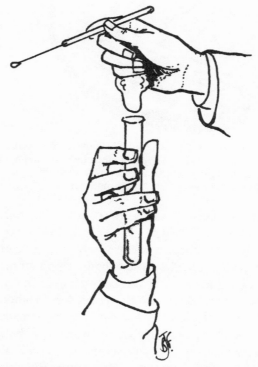

Fig 22a: Collecting a sample. The standard means of collecting a sample for culture has been little changed for the best part of a century. A test-tube of the specimen liquid, sealed with a plug of non-absorbent cotton wool, is cautiously opened so that contaminating extraneous organisms from the surrounding air have only a small chance of getting in. A sterilised loop of platinum wire (or nichrome, which is cheaper) is used to extract a droplet of the sample.

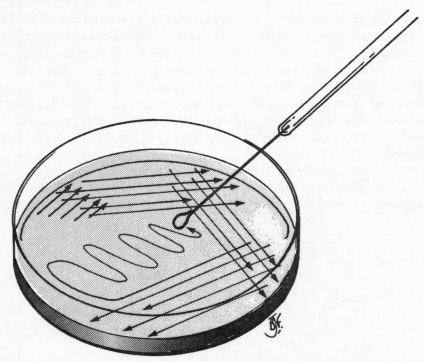

Fig 22b: Plating the sample. The wire loop is then streaked across an agar medium surface in a petri dish, following a pattern like the one indicated by the arrows. In this way successively infrequent bacteria are deposited on the agar, so that separate colonies are sure to appear somewhere on the medium.

media made with gelatine melt and become useless. Not only that, but some bacteria produce enzymes which dissolve gelatine. Agar is resistant to enzyme attack, and it has the unusual property of melting only near boiling point (100°C) though it solidifies close to body temperature. This fortuitous phenomenon means that agar gels will not melt even in a warm incubator, once they have set firmly. But it also allows one to add blood to the medium at around body temperature, which some pathogens demand. Melting point and solidifying point are almost always assumed to be the same: how fortunate it is for bacteriology that agar is a mystifying exception to that rule.

Agar is poured into screw-topped bottles, culture tubes or petri dishes for routine purposes, and when the medium is solidified and its surface dry, the bacteria are introduced on a platinum wire loop which has previously been sterilized in a bunsen burner flame. The addition to the medium of particular substances to inhibit or potentiate the growth of particular microbes enables us to culture one kind of organism out of a mixed batch; or alternatively when all the colonies have appeared one can pick out the particular type desired for study and spread that with the loop across the surface of a fresh culture vessel. It is in this way that pure cultures can be obtained.

Though bacteria most usually proliferate by division, they can also undertake a form of sexual reproduction. The first evidence for this phenomenon appeared in the 1920s, when experiments were being carried out with *Streptococcus pneumoniae*. Normal colonies of this organism are rounded and smooth, but it was noticed that occasional rough-surfaced colonies appeared on the culture plates, too. These new bacteria were dubbed R-strain (for 'rough') in contrast to the normal, smooth S-strain. Further culturing showed that the R colonies always reproduced true to type: they never reverted to the S-form. But if the bacteria were injected into an experimental animal then all that could be recovered were S-forms. The research workers were faced with the conclusion that:

(1) R-form bacteria cultured artificially did not revert to S-forms; but
(2) R-form bacteria inoculated into an animal could only be recovered in the S-form.

As part of the follow-up, a range of different combinations of the two strains were studied, and one of these experiments revealed a most unexpected result. Some of the R-form bacteria were injected in combination with some heat-killed bacteria which belonged to a different S-form. What was recovered from the animal was the S-form organism *which had been injected after being killed*. It was as though the dead cells had been miraculously resurrected – but what had happened was that the living R-form organisms had picked up some of the genetic characteristics of the dead S cells. So, for the first time, it was discovered that genetic material could be passed from one bacterium to another. Twenty years after that discovery, during the Second World War, the identity of the transferred compound was revealed: it was DNA, which we now know to be the fundamental genetic store common to all forms of life. It is this material, deoxyribose nucleic acid, that underpins all our understanding of the genetic blueprint of cells and which has the rare distinction, for a chemical, of having a book named after its structure – *'The Double Helix'* by James Watson and Bernard Crick, published in 1968. The analysis of DNA and the unravelling of its configuration has been an important milestone in the field of genetics. Once again, it was a microbe which pointed the way to this new understanding.

Microbe Power

The discovery encouraged further work in the field, and in 1946 the first example of a 'crossbred' bacterium was observed in a laboratory culture vessel. The experiment involved the bacterium which inhabits our intestine, *Escherichia coli*. Normally it requires – among other things – a supply of four amino acids: threonine, leucine, biotin, and methionine. Special strains were isolated which could themselves synthesize two of these compounds, one strain could produce its own threonine and leucine while the other synthesized biotin and methionine. If either was inoculated onto a growth medium that lacked all four amino acids, then the bacteria died out. The surprise came when a mixture of the two strains was plated out on the deficient medium, for some colonies did appear. At first sight this seems to be impossible, and it transpired that a small percentage of the organisms were transferring genetic material from one cell to another, so that a few of them – only about one in a million – ended up with the genetic information to manufacture all four of the essential molecules.

This in itself does not prove that conjugation – physical union of the cells – was taking place. There are other possibilities: perhaps the genetic programme was free and able to float from one cell to another, or possibly a virus was carrying the vital fragment of DNA between the *Escherichia coli* cells. To eliminate these possibilities, a simple glass U-tube was constructed. The strains of the bacteria, growing in a nutrient broth, were introduced into the two limbs of the tube, and they were separated by a porous filter. The growth medium was pumped back and forth through the separating filter, while it kept the bacteria apart. No genetic changes occurred, so clearly the bacteria needed to come into physical contact for the transfer of genes to take place. It was the electron microscope which revealed what was happening. Individual *Escherichia coli* were lying touching each other. A projection could be seen to pass from one cell to meet with the partner bacterium, and it was through this that the DNA was being passed. Further research showed that one cell was donating genetic material to the other, and the two strains were referred to as 'male' and 'female'. It is a mechanism reminiscent of copulation in man – and an interesting example of parallel developments in dissimilar organisms.

Before becoming too carried away by this notion of microbes coupling in diminutive ecstasy, like lovers in the park, we have to see the differences from human sex, as well as the similarities. In man, as in other higher forms of life, sexual reproduction consists of the uniting of two sets of chromosomes, forming a fertilized zygote which has the same total of chromosomes (i.e. a complete diploid, or double, set) as the parents. And this is as true of ferns and fish as it is of birds, bees, and people. In bacterial conjugation this does not occur. A small part of the genetic material of the male strain is injected into the female, and that is all. So in these terms the processes are not directly homologous. However the occurrence of physical conjugation, the existence of a protrusion specifically designed to unite the male and female cells, and the passage of some genetic material, is exceedingly interesting, since it reminds us that bacteria can tell us a lot about the origins of multicellular organisms and shows that (for all their small size) these microbes are not quite as unsophisticated and lowly as they seem.

The transfer of genes from cell to cell even without sexual contact has been observed as well. The first indication of this unexpected phenomenon came in 1952, when the glass container described earlier (see p. 156) was used for a series of experiments – but on this occasion, some genetic charac-teristics were passed through the separating filter even though the bacteria themselves had certainly not been in contact. The spread of a character from one population of microbes to another like this is similiar in some ways to an infection, and this comparison proved to be justified for it was discovered that a virus had accidentally acquired part of the bacterial DNA – and the virus was able to pass through the filter. This property is now known to explain some of the instances of bacteria acquiring resistance to antibio-tics – they literally 'catch' their new property – and it may account for the way some microbes become pathogens. And of course, the use of enzymes to remove portions of DNA and transfer them from one cell to another artifi-cially underlies the technique of plasmid transfer, which can enable mic-robes to be given a specified new property. Already it has been shown that microbes which have never had the ability to fix nitrogen can be given that characteristic in the laboratory, for example, and the possibilities for further permutations on that basis are endless.

I have referred to viruses, and we must consider them briefly, although in the strict sense they do not belong in a book on microbes. Whereas microbes are single-celled organisms, viruses are not. Instead, they are minute biochemical complexes of genetic material which can take over part of a cell's machinery and use it to produce more virus. Although they are sometimes infectious and disease-causing, then, they are not *living* in any ordinary sense.

The word 'virus' derives from the Latin and means *poison*. For several centuries it was used spasmodically as a description of the infective principle of disease, and in the mass media today it is still sometimes used for this purpose. An example is a recent newspaper account that an outbreak of typhoid had been the result of an 'exceptionally virulent form of the virus'. Typhoid is caused by a bacterium, and not a virus at all; but before we condemn that use of the term as entirely inaccurate we should remember that it does have an historical precedent.

Microbes do not normally interfere with the growth of living cells, whereas viruses usually do so to a greater or lesser extent. While microbes reproduce by dividing, viruses do not reproduce at all. New virus particles are assembled by the host cell in response to the pirating of the cell's machinery by the invading viruses, a process which is separately known as replication. Thus, microbes in division bequeath all their cell substance to the succeeding generation; in contrast to this, viruses pass on only the genetic blueprint to instruct the production of their successors. All microbes contain DNA and RNA – the two types of genetic material – but viruses contain only one or the other. Further distinctions between microbes and viruses arise when we consider the way they live. Viruses do not contain enzymes as a rule (and the few exceptions to that rule boast only one or two). Microbes, on the other hand, produce a large range of enzymes that initiate and carry on the microbe's life chemistry in a way no virus can.

I have referred to half-a-dozen important points of contrast between microbes and viruses, but these different considerations can be further simplified, giving us two main points of distinction:

(1) No virus can ever arise directly from a pre-existing virus; but a microbe is always produced from an existing microbial cell.

(2) Viruses are incapable of life. By hijacking the cell machinery of the host they can be replicated; but they do not carry out the normal processes of living like cells.

Not only are viruses much smaller than living cells, they very often have a bizarre appearance. Some take the form of spiked, sculptured spheres. Others are spirally wound rough cylinders. Many are faceted, like elaborate baubles from a Christmas tree; and some phage viruses which infect bacteria look almost like a space probe. Their contracting stalks, hexagonal base, and the crystal-shaped 'head' are unique; indeed they function like a miniaturized injector, undertaking an arduous free-floating journey from one cell to another and equipped only to inject the genetic messenger which causes the cycle to be repeated. The analogy with a spacecraft is not only appropriate because of their appearance, then; they also behave just like space probes.

Viruses are usually so simple that you can make molecular models of their structure by glueing table-tennis balls together. One tacitly assumes that these models are essentially artificial: a handy means of visualizing their composition, and nothing more. But viruses really do look like that, and the electron microscope shows that the molecular models give an excellent impression of what the virus would look like to its host.

It is possible to crystallize viruses by allowing solutions of them to gently evaporate, leaving behind crystals of the solids like those of many other chemicals. In some respects it is most convenient to think of the virus as a fragment of freelance genetic matter, unable to reproduce and interfering with a suitable host cell like a collection of nomadic genes. This explains the difficulty attached to finding a cure for virus diseases. Since they become so closely involved with the host cell, anything that is likely to incapacitate the virus will very likely kill the host at the same time. Though there are many vaccines available which can prevent virus diseases, there is no successful drug which can cure a virus disease. It is often said that science has yet to find a cure for the common cold – but this is just one group of virus diseases out of many and we ought to bear in mind that the same failure to develop a cure applies to every other virus-induced illness. The few antibiotic substances which have shown some tendency to act against viruses have not been generally useful, the development of the natural antiviral substance interferon has failed to materialize in practical terms, and an outbreak of virus disease could decimate mankind. One can only hope that a 'supervirus' does not emerge or, if it does, that by then we have found an antiviral drug from a microbe which can cure the disease successfully.

Viruses have an intimate relationship with the world as we know it. The broken red and yellow colour of tulips is caused by a virus infection, for example. The curl of the potato leaf is apparently due to a chronic virus infection, too. It is only now that modern culture techniques have enabled us to see the flat, healthy leaf of an uninfected potato plant for the first time, that we know what it should actually look like. Might it not be possible that some other innate characteristics of living things are caused by viruses? If this is so, then certain genetic features might be passed on infectively. The notion of evolution by contagion would have far-reaching consequences, and even the more immediate prospects for pest control through the use of virus sprays could revolutionize agriculture.

Finally, one word of caution. In many circles it has been popular to think of viruses as the most primitive 'organisms', and a good many teaching books have shown the virus as the basic stock from which more complex living things have arisen. When we think of viruses as simplistic replicating systems it is easy enough to conceive of them as the progenitors of life. But when we realize that they are genetic entities, and not microbes, and that they rely on the host cell for their self-perpetuation, it is clear that they can never have been independent, nor could they have existed in a world devoid of higher living things.

If viruses could not exist until there were already available cells to undertake their replication, where did the first life come from? The hypothesis that has gained credence over the past few decades has been affectionately christened 'the primeval soup theory'. It suggests that life arose in the complex brew of organic raw materials that were present in pools of shallow water when the earth was young. The principle evidence in favour of this view is the production of amino acids and other raw materials of life under laboratory conditions. A blend of the compounds that may have existed in the young earth model are mixed with water in a flask, warmed, and treated to a variety of energetic phenomena including ultraviolet illumination and electrical discharges in imitation of lightning.

The end result is that traces of organized molecules (including those that could be the precursors of simple proteins) can be found in the 'soup'. This does not mean that proteins are going to appear miraculously in the brew, nor does it imply that we know how to simulate the conditions that actually existed when the earth formed, so it is not a hard-and-fast proposition. Indeed, it is possible that such an environment might serve to break down more organized molecules as much as to build them up.

There is another problem, too. Because this theory fits so many of the observable data, and confirms our assumptions, it tends to make us feel that it is likely to be the answer simply because it had diverted attention from possible alternatives. I think we may yet find that life arose from compounds that were generated in outer space, and not on the earth's surface at all. In recent years there have been several reports of amino acids being found in meteorites, which has raised the possibility that they have originated from forms of life elsewhere in the universe. A range of organic molecules has been detected by radio astronomers, which seems to support that view. However, it could allow us to reach a more intriguing conclusion – that these compounds can arise in space through chemical processes. I have said how complex precursors of life on earth might have been broken down by chemical interactions – but in space these objections would not necessarily apply. Molecules *en masse* cannot oxidize successfully in the vacuum of space, nor can they hydrolyse with just a few molecules of water. But the demonstration of organic chemicals in gas clouds shows us that the synthesis can certainly occur. Perhaps life arose in space, or if not life (which would be carrying the argument too far), then at least the complex molecules which started it off.

Consider space, rather than the 'soup', as an alternative. Here we have a range of available elements, particularly the atoms that go to make up living systems – hydrogen, carbon, and the rest – and there is ample energy in the form of radiation that might initiate the reactions. The molecules that have been found in space include many (such as water, alcohol, and carbon dioxide) which are associated with life, and in late 1972 it was reported that a complex spiral molecule had been found in a meteorite. Yet this was not the analogue of a kind of 'DNA' left over by some alien form of life: it seems that the molecule formed chemically.

We have already touched on the fact that complex molecules can exist in mirror-image versions of each other, so that we have D-(right-handed) and L- (left-handed) versions of the same compound. What is so surprising about the spiral molecule from the meteorite is that the molecule was symmetrical, and did not conform particularly to either the left- or right-handed configuration. The only likely origin for an exactly balanced product like this is a chemical reaction, so we can suggest that it might have arisen chemically in outer space. If a molecule as complex as that can arise under such conditions, it is possible that the raw materials for life came to earth partly assembled and ready for action. The 'soup' may have given rise to its first living occupants not as a result of earth evolution alone, but by providing a suitable substrate for the further development of space life. The evidence now accruing is certainly compatible with this thesis, yet what an intriguing prospect it is. One earlier theory on the origin of life on earth (the 'panspermian' hypothesis) held that life was once disseminated throughout the universe in the form of resistant spores. That view says nothing about the way the life originated in the first instance. But the 'space-life' approach could account for the origin of some of the raw materials for life, and it argues in favour of the widespread dissemination of them. If so, the prospect for life on other hospitable planetary systems could be realistic.

8 The Microbe Attitude

I would like to describe a remarkable little microbe. It is shaped like a flattened pear, and swims – pointed end foremost – by means of a long and tenuous flagellum which thrashes about, so that the entire cell rotates regularly as it steadily spirals along. What is it? From the description it sounds like a protozoan, a free-swimming, independent animal cell.

It is a man.

This free cell, swimming purposefully along, is a sperm. As a protozoan, it would live in water throughout its life, forming a hard cyst wall to protect itself if its watery environment dried up. The sperm has a different way around the same problem. At the end of its swimming stage, it unites with a larger cell, known as the egg; and from the resulting fused cell grows a huge colony of diverse cells, some of them light sensitive, others responding to sound; some secreting active chemicals such as enzymes, others acting more or less passively as structural elements. This elaborate mass of cells is a fruiting body known as a 'person'. In it are formed the single-celled protozoan-like organisms we began with.

So man is a single cell: the people we see around us and in the mirror are merely the fruiting bodies which produce it. All the genetic information required to make the complexity of an adult is contained in the cell, the head of a sperm; this must be the most condensed form of information yet devised – truly the ultimate microdot.

This relationship between the most highly evolved multicellular creatures – ourselves – and the microscopic single cell acting independently, spans the entire length of evolutionary progress. It reminds us that for all our structural complexity, it is in the form of single isolated cells that we are passed from one generation on to the next. In procreation man reverts to a microbe. The purpose of the human body is, in functional terms, nothing more than an elaborate means of ensuring the continuity of this germ line, by providing the right milieu for the cells of procreation to survive. If the sperm was a protozoan, it would have its lake in which to swim; but since evolution allowed us to leave a watery environment there is the necessity to maintain an artificial, internal lake instead. The organs of sense are primarily there to obtain water, and to obtain food to nourish the supportive colony of cells. The sexual apparatus is a simple mechanical means of ensuring that the continuity of the fluid environment is not broken.

In flowering plants there is a similar return to the ancient origins of today's species. Pollen grains are no more than spores. When they alight on the pistil and germinate, they look under the microscope exactly like a fungus spore sending out its first hypha. In the higher plant, this fine cell is called the pollen tube, and it eventually fuses with the ova to start off the chain of events which results in the setting of seed. But the independent stage, that free pollen grain, is a spore and in its brief life-span it behaves just as one would expect such a microbe to behave.

This different way of thinking about higher organisms can give a new slant to many of the philosophical difficulties of life, and how living organisms are organized. In this way we can begin to talk about microbe awareness as an attitude of mind. For example, we were all taught that the essential items of our diet nourish different parts of our bodies – calcium is for the bones, carrots help one see in the dark, and so forth – and this is the basis of what I call the *pigeon hole theory*. Like sorting letters into pigeon holes, the idea seems to be that elements in the food have a predestined target to which they miraculously home.

When we stop to consider man as a community of cells, however, a clearer alternative picture becomes apparent. We can now see what effect a given dietary constituent has on the individual cell's behaviour. Thus, ascorbic acid, vitamin C, is involved in the production of the cement substance that binds cells together. Its role in the formation of collagen, the fibrous protein that gives animal tissues their resilience, is vital. This is all we have to know to understand what happens when a diet contains too little of the vitamin. Sores develop, the gums become raw and inflamed as the teeth loosen, there may be signs of a degeneration in the structure of heart muscle, and there is pronounced weakness. The student of pathology or dietetics has to memorize those symptoms as a catalogue of woes; and this is the parrot-fashion memorization without very much understanding which the pigeon-hole theory engenders.

If, on the other hand, we picture the effects on the cell of too little cement

163

substance holding the cells together, all the symptoms become necessary consequences of that fundamental deficiency. If the cells are not binding together as they should, you would expect the teeth to loosen, the gums to become raw, and blood vessels to rupture, and so on. The apparently unrelated clinical signs of scurvy become immediately comprehensible, and relate to a single disturbance at the cellular level. When we learn, in addition, that lack of ascorbic acid interferes with other aspects of metabolism – particularly the uptake of folic acid, one of the B vitamins vital for the formation of blood cells – then the anaemia typical of scurvy becomes understandable, too.

This way of looking at the health of a whole organism by considering one of its component cells can be extended elsewhere. Many people have objected to the use of animals in research, and although in many instances their use is vital (and in the security of a warm animal-house, it is a safer and better-fed life than the majority of domestic pets can have), there is a strong case to be made out for the use of a single cell as an experimental animal instead. Whole animals are utilized because the effects of a drug or a form of treatment on the functioning of an anatomically complex creature cannot be imitated by using organs, or tissue cultures. But the modification of the behaviour of a single cell – if we use a cell as a microbe-like equivalent of the whole animal – could provide some useful clues. Indeed I have seen, during the compilation of this final chapter, a paper which suggests a way of doing just this with one of the white cells found in the human blood stream, and entitled 'The human lymphocyte as an experimental animal'. Studies of a single living cell could tell us much about the functioning of the entire community in the form of a living animal.

This analogy between a multicellular animal (like ourselves) and a community of single microbial cells can give a novel interpretation of many aspects of biological philosophy yet, though it was a view which began to gain currency when the existence of cells was first discovered in the middle years of the nineteenth century, science has tended to lose sight of it since. Is it a valid idea? Well, the ultimate test is whether a single human cell can really act like a microbe, and it certainly can. The easiest example of that is the scavenging white cells in the blood stream, which live and behave just like amoebae and crawl around in among the cells of the blood and the areas where there are mopping-up operations to be done, following injury or infection.

But most graphic of all, in my view, is what is happening to the cells of an American citizen in countless laboratories spread right over the world. Cells from this women have spread through tissue cultures just like a microbe contaminant would do. As a result, many cultures are being literally taken over by the cells – yet, most surprising of all, the women from whom they originated died many years ago!

It began when the woman, a negress named Henrietta Lacks came into the Johns Hopkins Hospital in Baltimore in February 1951. She was being treated for cancer of the cervix, and as part of the investigations an attempt was made to culture her cells from the growth. The cells took well to an environment of serum and proliferated in the test-tube as well as they did in her body. In due course they were subcultured to fresh culture bottles, and in this way the cell line became established. Mrs Lacks' treatment, in those earlier days, was not as successful as it might have been today, and she eventually died of her condition.

Her cells, however, lived on. Under the abbreviation of HeLa cells (the first two letters from her first and last name) they were successfully subcultured, and have been ever since. I dare say there are many tons of the cells of Mrs Lacks alive in laboratories around the world, each one bearing the potential to recreate the entire human from whom they came. It is important to remember that a living cell's nucleus contains all the genetic information that was present in the original fertilized egg cell, since the dividing chromosomes produce an identical new set at each cell division. In many species a single cell within the adult divides as though it was a zygote (a fertilized egg cell), to give rise to an offspring. This is the way in which vegetatively reproducing plants give rise to bulbils, which are small, new plants sprouting from the parent's stem or leaves. A similar phenomenon occurs in animal species which reproduce parthenogenetically – through 'virgin birth', that is – such as aphids (greenfly) and the silverfish insect. The principle is generally known as *cloning*.

The fact that Mrs Lacks' tissues survive in huge quantities long after she is dead poses some interesting food for thought about biological theory, quite apart from its almost eerie reality. First, it makes us realize the importance of cloning as a process for, if it was a phenomenon that could be easily induced to occur, then we would have no difficulty (in theory at least) in trying to revive Henrietta Lacks. In the same way we could easily plan to reincarnate the elderly as artificially implanted foetuses by cloning cells from their bodies and allowing them to be born anew. But we cannot do this. It may be that the body cells, the somatic cells as they are called, are descended from 'master' cells in the tissues through a mechanism that is still obscure. If this kind of mechanism exists, or something like it, then there could be a fundamental inadequacy in the cells that meant they could never be cloned. Either way, a study of the individual cells from a multicellular body can tell us much about these important, unresolved issues.

A second area of insight that HeLa cells (like cultures of other cancer cells) can offer is the prospect for immortality. Mrs Lacks, had she not died of cancer, would certainly have died of old age – but her cells have not. They grow and reproduce incessantly, without showing signs of ageing like the cell community which was her body. Here is another interesting comparison with the microbe world, for it is single-celled organisms that show this potential for virtual immortality in nature. Perhaps the most vivid illustration of the way in which these cultured human cells can behave like single protozoa is the way they have accidentally become added to cultures of other types of cells, and in time have come to predominate. In just the same way that an unwanted species of microbe can contaminate a culture vessel, HeLa cells have recently begun to turn up in cultures where they are not wanted.

It is strange to think that the research worker who first tried to culture the cells in 1951 – George Gey – was excited and no doubt surprised when he succeeded, for this was the first time that any human tumour cells had survived in culture. Since then they have proved to be uniquely adapted to life in the test-tube, and they seem to be more vigorous than any of the other human cell lines routinely cultured. Just one HeLa cell introduced into a culture of something else (through the use of a contaminated pipette, for instance) is enough to finish the culture, for in time the HeLa cells will take over.

It now seems that there is far more HeLa around than most research workers know, for the National Cancer Program in the United States has begun a survey of hundreds of different cultures from laboratories all over the world – cultures of tumour cells ranging from breast to bladder – only to find that the cells are not what they seem. When detailed tests are carried out, only too often the result is that the culture is of HeLa cells, and not the original cell line at all. The most outstanding example of this is probably the exchange of cultures which the Americans and Russians undertook in 1972. In both countries, virologists had obtained evidence of viruses in certain cancer-tissue cultures. This does not mean that the viruses caused the cancers, of course; they could have been present coincidentally, perhaps they were released in large numbers as a result of the transformation to the malignant state, or possibly they were present purely as a contaminant of the culture – there are many possibilities.

To find out more, virologists in different laboratories exchanged cultures in order to test out their own methods on other people's material. Altogether, six different kinds of tumour cell were sent from the Soviet Union to the United States. But detailed work began to suggest that the cultures were very similar – even identical. The cells had a lot in common with HeLa cultures, and indeed most of the research workers were quickly convinced that they were HeLa cells. This was certainly possible, since cultures of HeLa had been sent to the Russian scientists many years earlier (long before the official exchange of material was begun). The final clue came from an analysis of some of the enzymes within the cells, particularly the enzyme glucose-6-phosphate dehydrogenase (known as G6PD for short). In Caucasians, this enzyme occurs in a variant know as 'type B'; in Negroes (and those of part Negro stock) the enzyme occurs as 'type A'. Mrs Lacks was a Negro – and the Russian cell cultures were found to have the type A G6PD, indicating Negroid origins. More topical is the suggestion that many other kinds of cell culture, which were being used to study possible differences between one cell line and another, may actually be HeLa contaminants. Under the microscope, they all look very similar when growing in a culture vessel and it takes detailed chromosome counts and several other kinds of test to tell HeLa cells from the others.

The problem for the specialist, then, is considerable. He is beginning to realize that he cannot trust his cell line any more – and he may be wondering whether the vigour of the HeLa cultures is such a good thing after all. As far as our attitude to microbes is concerned, this continuing mystery is a textbook example of the way in which human cells, under the right circumstances, act as though they were independent microbes. And lurking in the shadows behind all this is the implication that the transformed, malignant HeLa cell – able as it is to exist like a microbe, without ageing – has acquired some new properties in the process of becoming transformed. One tends to think of cancer cells as being cells that are in some way lesser cells than normal, that are rogue cells hell-bent on destruction. That should not disguise the fact that these cells in culture have some distinct advantages over somatic cells (i.e. cells in the body). To confound this still more, we know that agents which transform cells are usually agents which damage cells: so we now compound the anomaly. 'Damage' a cell and you would expect it to become less of a cell; but here we end up with a cell that has acquired immortality of a sort, and which outlasts its 'undamaged' counterpart.

My answer to that as a student was always to argue that the effect of the damage was to interfere with those regions of the cell which regulated and controlled its growth. In this way a cancer cell, at the moment of its transformation, became 'less' of a cell in that its growth-regulating, 'braking' system was less than it had been. The cell, in consequence, was released from restriction and continued to grow – which is, in simple terms, what happens in cancer.

This alternative view of malignant transformation could best be illustrated by the analogy of a self-powered car. If this is our 'cell', then it is difficult to imagine how it could actually drive faster by being damaged. One would expect that a car in a damaged state would tend to go slower, or to drive less well, than it had done before. But that only applies if we regard the car as being accelerated by stepping on the gas. Suppose instead we design a different type of vehicle, in which the accelerator pedal is fixed to the floor and the car's speed is controlled by the brakes, instead. Then it becomes easier to see how a damaging effect could make the car go faster: if the braking system was damaged, then a runaway increase in speed is exactly what one would expect. Cancer cells do not always reproduce faster than normal somatic cells, incidentally; what distinguishes them from organized tissues is that they continue to divide long after they should have stopped (and apparently keep doing so almost indefinitely), and secondly that they invade other parts of the body and spread far and wide, disseminating the tumour itself in the process. The runaway cell idea explains the essence of cancer in an easy-to-assimilate form; and it underlines the way in which the most structurally advanced form of life – mankind – is little more than a colony of microbes for whom organization and control is of paramount importance.

How does this structural complexity arise? It has often been argued in the past, by philosophers in biology, that the relationship of a single cell to a complex multicellular organism is a difficult one to resolve. The biologist Paul Weiss has put it thus:

'At the beginning of development we find just one primordial cell – the egg. We call a system 'organised' when its multiple elements appear in typical diversity . . . In the developed system, 'organism', the cells represent the elements; hence, organisation is a supra-cellular property. But the primordium of the organism – the egg – does not consist of cells. Now, there arises a dilemma.'

In other words, either the egg already possesses the specialized parts that living organisms need (in which case it is not a cell at all, but merely a body not yet divided up into cells); or alternatively it is nothing more than a single cell, like all the others in the mature organism, in which case it could not be a structurally organized living thing like the whole organism.

The point is an important one, and in one form or another it has been around since 1838, when Theodore Schwann published his *Microscopical Researches* which showed how the cell was the universal component of living organisms. The philosophical debate as to which view is correct has been conducted at great length, and in exhaustive detail; but it seems to me that we might be put out of our misery if we realize that there are serious objections to both points of view. The 'egg equals cell' viewpoint cannot be right, since for the time it exists the fertilized egg is the entire organism, whereas a cell from an adult human is an infinitesimal part of the whole. Our microbe neighbours can give many examples of protozoa in which the cell is a complex and specialized structure, showing many of the features that one ordinarily finds only in many-celled creatures (the complex protozoan *Epidinium* is a suitable example), so it is clear that in philosophical terms the single cell can equal the entire organism. But the human egg is more than equivalent to one cell from the adult.

The alternative view that 'egg equals organism' is equally easy to fault. Though the egg is much more than just a single body cell, and has the potential to develop into a new, previously unseen, unique individual; and although some cells (like *Epidinium*) do develop the structural complexity of multicellular animals, none the less an egg cell is not a condensed version of an adult. It is not merely that the cell has yet to divide up into compartments, as it were; it has to generate countless individual cells which themselves take on the task. So, although it is perhaps closer to the truth than the 'egg equals cell' point of view, even the 'egg equals organism' hypothesis is unsatisfactory.

What we need is a single set of ideas which can link these notions into a unified understanding; a way of making a complete tapestry from the separate strands of evidence. The approach I envisage is a theory of delegation: a proposition that the many cells of a complex organism carry out singly the functions which are embodied in a microbe. These multifarious activities are selectively repressed as the cells of a multicellular organism become specialized. Thus a mature cell is able to carry out only those functions which are necessary for its particular role within the community; the other properties are repressed, even though they remain in coded, genetic form.

One would suggest, then, that the *Epidinium* cell reveals the full development of specialized functions with a single cell, while the vertebrate body displays a similar arrangement through the delegation of specific functions to cells produced in the temporal and anatomical order which give rise to the integrated choreography of many-celled life. The microbe produces 'nerve' fibrils and 'muscular' elements within itself: the vertebrate body, by analogy, programmes its cells in the regions predestined for nerve function to repress conflicting activities, channelling its capacity solely into the development of nerve material, and in the process becoming a neurone; and instructs those cells programmed to become muscle to develop the contractile muscular proteins above all else.

The specialized cells of the adult body, like any somatic cell that is to survive at all, naturally embody all the metabolic functions necessary for life. We will doubtless find nerve proteins in muscle cells, and traces of muscle protein in any cell likely to move, even slightly. But apart from these essential proclivities, the delegation concept allows us to explain why it is that a single cell (like a microbe) embodies so many of the functions which

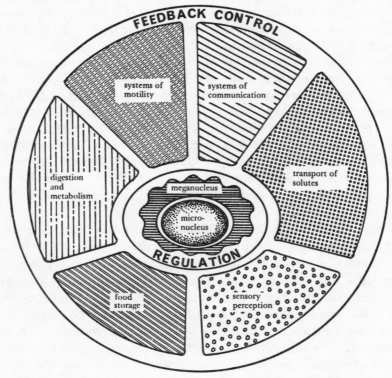

Diag 13 a: The single cell of a protozoan microbe. This simplified scheme would account for the way in which specific areas of a single cell develop separate functions; in contrast to the pattern seen in many-celled animals.

170

we find in multicellular creatures. The advantage of multicellularity is not merely an increase in complexity correlated with a similar increase in ability or anything else – we have seen that the 'simplest' protozoan can carry out many of the functions of higher animals, so a multicellular life-style is no guarantee of superiority in that vital sense. It is that the many-celled organism can produce specialized, independent structures for the control of disease, the repair of the whole, and the regulation of the inner milieu, whereas a single cell is inherently more vulnerable to trauma. Damage one cell in a man and it would be virtually impossible to tell: damage one cell of a microbe and it is wounded in entirety. The skin cells which provide us with our waterproof, protective coating are sacrificing themselves in the process

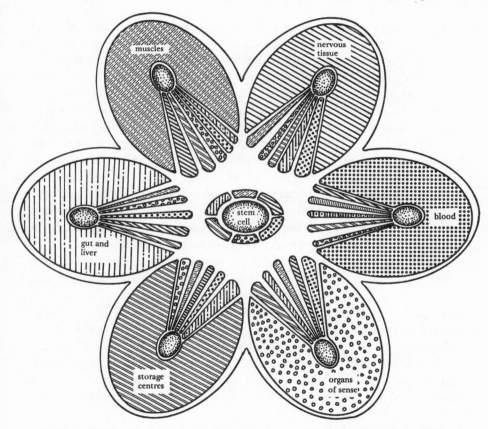

Diag 13 b: In many-celled animals, specific organs take over the functions confined to special regions within the single cell of a microbe. This diagram of a simplified animal shows how specialisation can be related to the way protozoa develop.

of producing a dead, resilient outer layer for the good of the cells – the man – within. In a sense, this is microbial altruism of the kind we met on p. 145, and a mechanism like this is founded on coordination of a uniquely sophisticated kind.

It is when the coordination is lost that we run into trouble. Many of the degenerative diseases in higher animals, and the illnesses in which the body seems to turn on itself, could be correlated with this view. Cells which lose contact with the role for which they have been produced can run amok: instead of removing extraneous, alien protein, for example, they could act against some of the body's own cellular components. A system evolved for the elimination of unwanted intruders could produce devastating effects if it turned against itself.

A cell which suffered damage could easily become a cancer cell, and according to this view, the destruction of centres which coded for the repression of its general behaviour, allowing it to become instead a specialized cell, would theoretically account for many kinds of malignancy. How interesting it is to look at the special characteristics of cancer cells: they are indeed less specialized than the typical somatic cell, just as one would predict. Their capacity to spread, to form isolated new colonies, and to exist in a state of steady growth without the normal tendency to age seen in the body, are all characteristics of an independent microbe. So we can say that as evolution proceeds, the delegation of cells to perform specialized functions becomes apparent; while in cancer the trend is reversed, latent tendencies for an independent life are derepressed and so revealed, and the corporate good of the community at large is lost altogether. No wonder Henrietta Lack's cells are now so like the microbes from which we all originated.

Does this idea bear any relation to the complex protozoa which we have already discussed? I think it does. The complex ciliates rely on their two nuclei, the micronucleus, which seems to act as the cell's code storehouse and which is involved in sexual reproduction, and the larger meganucleus, which runs the daily workings of the cell but vanishes at each sexual phase, only to be produced anew in the daughter-cells. It now turns out there may be a great many copies of the basic set of genes in each cell. The suggestion is that the micronucleus holds the 'master set', and the meganucleus has a number of separate copies run off at speed, and responsible for the various functions of the cell as a whole.

Those extra copies (there may be hundreds of them) may be the equivalent of the delegated cells of higher organisms, but in this case we have copies within the one cell of the microbe, rather than in whole sub-communities of cells as in animals like ourselves. In any event, it is possible to trace similarities between free-living microbes and the separate cells that make up our own bodies; and the insight these models provide can show us something of the way in which the human body functions, in health and in disease, and give a new look to the orthodox interpretations of evolutionary progress.

It can give us an insight into human death, too. The debate over the 'moment of death' has never been adequately resolved, yet for legal as well as medical or humanitarian purposes it is an important matter to settle. Death occurs only after the body's functional coherence has broken down, so that the interdependent physiological systems become incapable of mutual support. But to the cells, death means something less easy to quantify. Cells can run at low levels of turnover, if the circumstances compel them to; and it is entirely possible for a man to be 'dead' while virtually all his cells are alive. Take, as an example, a guillotine victim. At the moment that his head is severed, he is dead. But (apart from cells killed by the trauma of the decapitation) virtually every cell of his body is still alive *after* that time. The cells of which that human body is composed will died off when they run short of oxygen, so that they undergo cellular suffocation, unless they were poisoned by accumulating waste products first. That head, lying in the basket, has the look of death about it in your mind's eye, since broken bodies of this sort are dead bodies – by definition.

To the guillotine victim, however, the reality may have been very different. There is nothing to suddenly kill the cells in that severed head. I see no reason why the brain should not have continued functioning for a time afterwards, in which case the notion that decapitation is a quick and painless death would be very far from reality . . . the victim would know perfectly well what happened, and would remain consciously aware of his fate for quite some time after the execution had occurred.

The way in which groups of cells survive in the body after death is well enough known from the continuation of beard and hair growth in a corpse, and the occasional reflex movements in muscles that may occur hours after death. The use of respirators and the concept of intensive care, which has become increasingly widespread in the medical arena over the last couple of decades, has shown how often the entire cell community of a human body can be maintained in a coherent state even after the brain itself has ceased to function. But this does not solve the problem of 'when to switch off the machine', because the cessation of brain function need not be permanent. In some cases an unexpected revival can take place after months, when it is least expected. One such example is an American woman, Carol Rogerman, who at the age of nineteen had been in a coma for four months following a motor accident, and had wasted away to less than 30 kilogrammes in weight, when her previously inactive brain dramatically returned to consciousness. Only a matter of days before that, her parents had been advised to 'pull out the plug'.

A microscopic examination of the brain of one of these patients reveals that what clinicians talk of as brain death is difficult to reconcile with the state of the brain cells themselves since, very frequently, they are normal *and* alive. Perhaps the vegetable-like existence is due to a failure of the cells to communicate in some way, which is always a reversible option; but looking at brain death from the viewpoint of a single cell makes us realize that in many cases it is not 'death' at all. The implications of this view are far-reaching, since in a nation where medical care is costly (such as the United States) there are powerful economic realities to be faced in addition to the humanitarian considerations.

So what new criteria does the philosophy of single cells disclose? Firstly, it emphasizes that the death we write about and record is not the same as the moment of death within the cell community at large. My heart goes out to those cowboys in the movies who lie, gasping their last on the sandy road, only to flop back into the arms of the sweetheart or the sheriff, to be proclaimed 'dead' on the spot as their eyes close. All they might have suffered was a temporary black-out, heatstroke, shock, or a faint; but they were always buried under untidy mounds of dust and earth on Boot Hill or in the local graveyard before those signs of life returned.

More important than these fictitious accounts are the occasions when this kind of thing happens in wartime, in motoring accidents, or even in the home. Cases where a 'cadaver' revives in the necropsy room are rare, but still occur. Tests such as holding a feather under the nose or searching with a mirror for condensation from feeble breathing are still recommended. So when does the 'moment' occur?

The definitive instant does not, of course, exist. Death is a process of disintegration: a gradual sequence which begins with life and ends with the dignity of a return to chemical inertness. The cell concept makes us realize how futile it is to delineate a definitive time at which life expired. However (though this is doubtless sound enough as a statement of philosophy) we have to adopt some yardstick for legal and medical reasons. I think we must say that the moment of death can only be assessed retrospectively. Perhaps one person may died at an instant when another identical patient in intensive care unit might have survived; so an assessment of a patient's condition, on its own, is no guide in practical terms. Any definition of 'the moment of death' must take into account the place where it occurs, and must accommodate the likely development of new techniques of resuscitation.

Put like this, the definition becomes the moment at which the dying patient commences deterioration as the coordinated functions of the body cease to be mutually supportive, or the moment at which no medical expediency can restore them. To use a graphical analogy, it becomes *the time when the body temperature of a dying individual begins its steady fall from its maintained, metabolic level (normally around 37°C (98.4°F)) to the temperature of its surroundings*. In this way we can include cases (like the beheaded man) who is clearly dead, even though metabolically alive in every sense; it can encompass the patient who is resuscitated and artificially maintained by the expedient use of a respirator; it can apply to the terminally comatose; it embraces victims of accidental or therapeutic hypothermia; and it allows us to anticipate a time when research has given us new hope for the revival of those who, in our present state of knowledge, are doomed to death. It removes the curiosity from those cases of cardiac arrest whom the newspapers bill as 'The Patient who Died Twelve Times' – and it even encompasses the unlikely development of surgery so dextrous and so quick that it enables one to suture that severed head back into place!

There may also be lessons in the same cellular approach for anaesthetists. For major surgery, such as the open-heart procedures routinely used for the repair or prosthetic replacement of inefficient heart valves, drugs are used which prevent the transmission of motor impulses – the signals which instruct voluntary muscles to contract. The reason for this is that a sudden reflex muscular movement could be hazardous to the patient, and the assault on the system that major surgery amounts to could otherwise bring about reactions in the form of a twitch or a spasmodic jerk. Once these motor impulses have been quelled, somewhat shallower levels of anaesthesia can be used, which allow the burden inflicted on the body by an operation to be kept to a minimum. During open-heart surgery, the pumping and gas-exchange functions of the heart and lungs are taken over by a by-pass machine which mechanically imitates both functions.

From the surgeon's point of view this is an admirable answer to the practical difficulties of such a major operation, and at first sight it is good for the patient. But go back to the cell once more. The anaesthetic agents carried by the blood bring about an insensible state. But while the patient is on by-pass, the anaesthetic gas is exchanged, along with the oxygen and carbon dioxide; and the levels of anaesthetic in the bloodstream can drop. A

175

brain cell in this situation begins to return to something near its normal state: and it is possible that a partial return to awareness is the result. The immobilized body, which is still under the influence of the 'paralysing' drugs injected earlier, will not show any response; but the cells involved in sensation and memory could clearly be at work. In recent years there have been a few reports of psychiatric abnormalities in such patients which can be traced back to the time of this kind of operation, and I find it difficult to exclude the possibility that it is the partial functioning of the community of cells responsible for sensation which initiate this process. If so, then continuous administration of anaesthetic agents during that important by-pass phase (which can last as long as eight hours or more) might be important for the patient's subsequent well-being.

Adopting the concept of a single cell, functioning in relation to others, can teach us many things. It can encourage a rethinking of medical practice in many ways, and make us look again at a good many bodily functions. It can even make us restate what we mean by death. The philosophy of the microbe – the good microbe, that is – can show us much about the way we work, and can direct our attention towards some new ways of tackling many of the long-standing problems that are so urgently in need of an answer. The options and the permutations of the various possibilities are countless: and whether we choose to benefit from the fact or not, the microbe itself is busily shaping the world, preparing for our future, and giving us life.

Index